Developing Graph Comprehension

FRANCES R. CURCIO

ELEMENTARY AND MIDDLE SCHOOL ACTIVITIES

NATIONAL COUNCIL OF TEACHERS OF MATHEMATICS

Developing Graph Comprehension

Elementary and Middle School Activities

FRANCES R. CURCIO
Queens College
of the
City University of New York

NATIONAL COUNCIL OF TEACHERS OF MATHEMATICS

Copyright © 1989 by
THE NATIONAL COUNCIL OF TEACHERS OF MATHEMATICS, INC.
1906 Association Drive, Reston, Virginia 22091
All rights reserved

Library of Congress Cataloging-in-Publication Data

Curcio, Frances R.
   Developing graph comprehension.

   Bibliography: p.
   1. Graphic methods—Study and teaching (Elementary)
2. Graphic methods—Study and teaching (Secondary)
I. National Council of Teachers of Mathematics.
II. Title.
QA90.C87   1989   372.7   89-35205
ISBN 0-87353-277-5

The publications of the National Council of Teachers of Mathematics present a variety of viewpoints. The views expressed or implied in this publication, unless otherwise noted, should not be interpreted as official positions of the Council.

Printed in the United States of America

TABLE OF CONTENTS

	Page
Preface	vii
Graphs—What Are They and How Are They Used?	1
Levels of Graph Comprehension	5
Collecting, Organizing, and Analyzing Data	6
Constructing, Interpreting, and Writing about Graphs	8
Classroom Activities	9

Activity	Topic	Graph Form	Grade Level	Page
1	Hair Color	People Graph	K–2	12
2	Eye Color	People Graph	K–2	13
		Block Graph	K–2	
		Picture Graph	K–2	
3	Height	Life-sized Bar Graph	K–2, 3–4	16
4	Travel	Object Graph	K–2	17
		Object/Person Picture Graph	K–2, 3–4	
		Picture Graph	K–2, 3–4	
5	Favorite Color	Block Graph	K–2	19
		Picture Graph	K–2, 3–4	
		Bar Graph	3–4	
6	Favorite Pet	Block Graph	K–2	20
		Picture Graph	K–2, 3–4	
		Bar Graph	3–4	
7	Favorite Ice Cream Flavors	Block Graph	K–2	21
		Picture Graph	K–2, 3–4	
		Bar Graph	3–4	
8	Favorite Color	Bar Graph	3–4, 5–6	22
9	Favorite Pet	Picture Graph	K–2, 3–4	24
		Bar Graph	3–4	
		Double Bar Graph	3–4, 5–6	
10	Number of Children in Your Family	Picture Graph (1-to-1)	3–4, 5–6	26
		Picture Graph (2-to-1)	5–6	

Activity	Topic	Graph Form	Grade Level	
11	Favorite Ice Cream Flavor	Picture Graph (1-to-1)	3–4, 5–6	28
		Picture Graph (2-to-1)	5–6	
		Bar Graph	5–6, 7–8	
		Double Bar Graph	5–6	
12	Favorite Game	Bar Graph	5–6	31
		Double Bar Graph	5–6	
13	Favorite Leisure Activity	Double Bar Graph	5–6, 7–8	32
14	Rooms with a TV	Bar Graph	5–6, 7–8	34
15	Height	Bar Graph	5–6, 7–8	35
16	Daily Schedule of Activities	Circle Graph	5–6, 7–8	36
17	Spending Daily Allowance	Circle Graph	7–8	38
18	Types of Dwellings	Bar Graph	5–6, 7–8	39
		Circle Graph	7–8	
19	Height and Standing Long Jump	Double Bar Graph	5–6, 7–8	41
20	Height and Foot Length	Double Bar Graph	5–6, 7–8	43
21	Raisin Experiment	Bar Graph	5–6	44
		Line Plot	5–6, 7–8	
		Stem-and-Leaf Plot	5–6, 7–8	
		Box Plot	7–8	
		Back-to-Back Stem-and-Leaf Plot	7–8	
		Double Bar Graph	5–6, 7–8	
		Multiple Box Plot	7–8	
22	Height of Plant over a Period of Time	Line Graph	5–6, 7–8	49
23	Height over Time	Line Graph	5–6, 7–8	51
24	Daily Temperature	Line Graph	5–6, 7–8	52
		Multiple Line Graph	7–8	
25	Time of Sunrise and/or Sunset	Line Graph	5–6, 7–8	54
		Multiple Line Graph	7–8	

References   57

Software   57

Selected Bibliography   57

Appendixes

1 A List of Graph Topics Appropriate for Different Grade Levels   59

2	How to Make Reusable Teaching Aids	**63**
3	Picture Labels for Object and Picture Graphs	**64**
4	Large-Box Graph Paper	**65**
5	1-cm Graph Paper	**66**
6	¼″ Graph Paper	**67**
7	5-mm Graph Paper	**68**
8	Daily Activities Data Collection Sheet and 24-Section Circle Graph Outline	**69**
9	Supplemental Graph Reading Activities and Answer Key	**71**
10	Sample Data Collection Sheet	**79**
11	Sample Height Data Collection Sheet	**80**
12	Raisin Experiment Activity Sheet	**81**
13	Height-over-Time Data Collection Sheet	**82**
14	Temperature Data Collection Sheet	**83**
15	Sunrise/Sunset Data Collection Sheet	**84**
16	Picture Graph Activity Sheet	**85**

PREFACE

Skill in the critical reading of data, which is a component of quantitative literacy, is becoming a necessity in our highly technological society. In particular, processing information presented in newspapers, magazines, commercial reports, and on television is dependent on a reader's ability to comprehend graphs.

To meet the needs of society, industry, and business, our students must become adept at processing information. As stated in the *Curriculum and Evaluation Standards for School Mathematics,* children must be involved in collecting, organizing, and describing data. They should be able to construct, read, and interpret graphs as well as analyze trends and predict from the data (NCTM 1989, pp. 54, 105).

This book is intended to provide elementary and middle school teachers and teacher educators with practical ideas on incorporating the graph-reading component of quantitative literacy into the instructional program. It can be used to supplement the teachers' editions of K–8 textbooks or as an elementary methods text for preservice and in-service teachers. It provides many suggestions for activities that can be used with youngsters in both small-group and large-group instruction.

In support of the *Standards* (NCTM 1989), the activities presented in this book provide teachers with ideas to emphasize exploration, investigation, reasoning, and communication in mathematics. Furthermore, suggestions for using the computer as a tool are presented in many activities.

This material can be used at different grade levels, depending on the learners' prior experiences with collecting and analyzing data. The data generated and collected by the students should be interesting and meaningful to them.

The results of a recent research study (Curcio 1987) and suggestions of others (Landwehr and Watkins 1986; Nuffield Foundation 1967; Russell 1988) indicate that elementary and middle school students should be actively involved in collecting real-world data to construct their own simple graphs. They should be encouraged to generate questions about the data (What do the data tell us? What don't the data tell us?) and to verbalize the relationships and patterns observed among the data (e.g., larger than, twice as big as, continuously increasing). In this way, the application of mathematics to the real world may enhance students' concept development and build and expand the relevant mathematics schemata needed to comprehend the implicit mathematical relationships expressed in graphs.

The development of graph comprehension skills is not meant to be isolated from the rest of the curriculum. Ideas are provided for general skills development that can be incorporated across the curriculum. Graph reading is not limited to the study of mathematics. Graphs are found in elementary and middle school science and social studies. Although in other disciplines graphical representation includes graphs, maps, pictures, and diagrams that may or may not include numerical information, only graphs that present quantitative information will be discussed in this book.

There are five sections in this book: (1) Graphs—What Are They and How Are They Used? (2) Levels of Graph Comprehension; (3) Collecting, Organizing, and Analyzing Data; (4) Constructing, Interpreting, and Writing about Graphs; and (5) Classroom Activities. The major portion of this book consists of the classroom activities that were designed for immediate classroom use. Supplementary materials are given in the appendixes. These materials include topics appropriate for children in grades K–8, instructions for constructing reusable aids for teaching graphing skills, different sizes of graph paper, samples of data collection sheets, and supplemental graph-reading exercises.

Thanks are due to the many teachers and children who participated in the field-testing of these activities. In particular, I owe special thanks to Denis W. Moore, principal of PS 104 in Brooklyn, New

York, and the late David Brown, principal of IS 227—The Louis Armstrong Middle School in East Elmhurst, New York. Sincere gratitude is extended to the Christopoulos family for assisting in the development of many of the materials used during the field-testing.

The NCTM reviewers who made helpful comments and suggestions for improving this manuscript during its stages of development are gratefully acknowledged. In particular, special thanks to James Bruni, Christian Hirsch, Bonnie Litwiller, William Masalski, and Albert Shulte.

GRAPHS—WHAT ARE THEY AND HOW ARE THEY USED?

Since the dawn of civilization, pictorial representations and other symbols have been used to record numbers of humans, animals, and inanimate objects on skins, slabs, sticks of wood, and the walls of caves. We can infer the usefulness of this method of recording and communicating information from an ancient Chinese proverb that states, "A picture is worth ten thousand words." When advanced written forms of communication were undeveloped, people relied on pictorial representations and symbols as a means of recording simple statistics. This method has proved to be such an efficient and effective mode of recording data that it is still used today, although quite modified.

The modern graph evolved from the work of the seventeenth-century philosopher and mathematician René Descartes. For the mathematician, the graph is "an invaluable aid in the solution of arithmetic and algebraic problems, the solution of mathematical formulas, and the representation of relationships" (Arkin and Colton 1940, p. 4). For the layperson, the graph is an aid for clarifying, organizing, and summarizing quantitative information found in newspapers, magazines, and advertisements.

Graphs provide a means for communicating and classifying data. Graphs allow for the comparison of data and display mathematical relationships that often cannot be easily recognized in numerical form. The traditional, most common forms of graphs found in newspapers, magazines, and advertisements are picture graphs, bar graphs, line graphs, and circle graphs.

More recently, some new plotting techniques have been recommended for inclusion in the curriculum by the Joint Committee on the Curriculum in Statistics and Probability of the American Statistical Association and the National Council of Teachers of Mathematics. These techniques include line plots, stem-and-leaf plots, and box plots.

In the following sections, traditional graph forms and some new plotting techniques are described. Suggestions for presenting each of these can be found in "Classroom Activities," beginning on page 9.

Traditional Graph Forms

Picture graphs. The picture graph (also called pictograph, pictogram, or pictorial graph) uses representative, uniform pictures to depict quantities of objects or people with respect to labeled axes. It is used when the data are discrete (i.e., noncontinuous). The ideographs or symbols used must be the same size and shape to avoid misleading the reader (Huff 1954).

In the early grades, children draw pictures or use photographs to construct graphs about their favorite colors, how they travel to school, and so on. Eventually, the pictures or drawings of real objects are replaced with uniform ideographs. The uniform ideographs may represent real objects (e.g., a stick figure to represent a person or a tooth symbol to represent number of teeth missing) or they may take the form of something more abstract (e.g., a triangle or square).

During the elementary grades, children encounter picture graphs both with and without a legend or key. Picture graphs without a legend are easier for children to understand because the ideograph and the item it represents are in a one-to-one correspondence. Once a legend is imposed, the ratio of each ideograph to the number of objects it represents must be taken into consideration when interpreting the graph. Fractional parts of ideographs (e.g., one-half of a picture) usually cause some difficulties for children. Ideas for introducing children to picture graphs can be found in Activities 4–7 and 9. More advanced picture graph ideas are in Activities 10 and 11. (See figs. 1a and 1b for examples of picture graphs without and with a legend, respectively.)

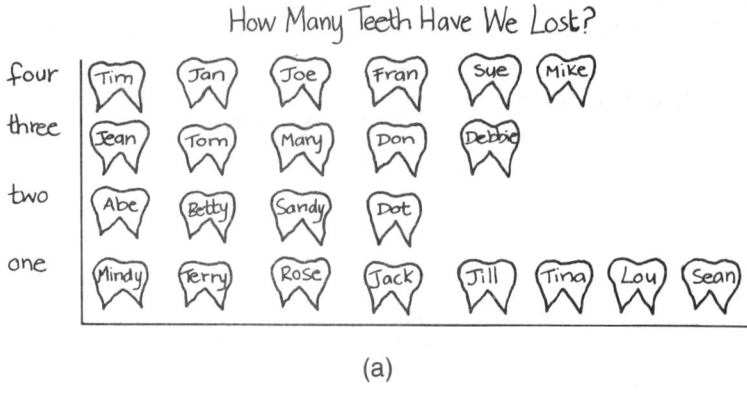

(a)

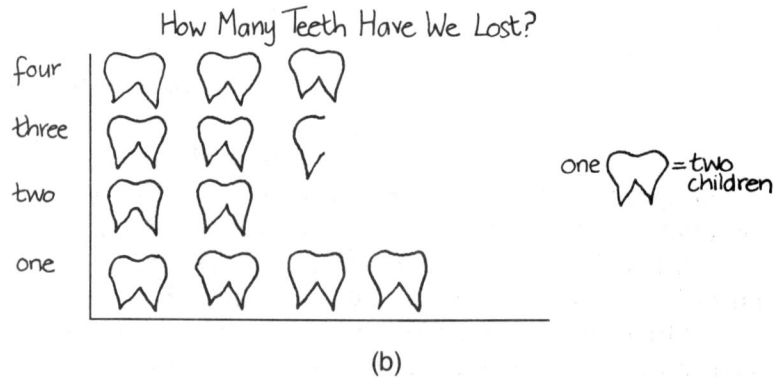

(b)

Fig. 1

Bar graphs. Set up horizontally or vertically, the bar graph (also called a bar chart) allows the reader to compare discrete quantities expressed by rectangular bars of uniform width, whose heights (or lengths) are proportional to the quantities they represent. The bars are constructed within perpendicular axes that intersect at a common reference point, usually zero. The axes are labeled.

Data presented in picture graphs are usually appropriate for bar graphs. Converting a picture graph to a bar graph is a natural way to help children move from a semiconcrete representation of data to a form that is more abstract.

Ideas for introducing bar graphs can be found in Activities 2, 3, and 5. More advanced bar graph ideas are in Activities 6–9, 11–15, and 18–21. See figure 2 for an example of a bar graph based on the data in figure 1.

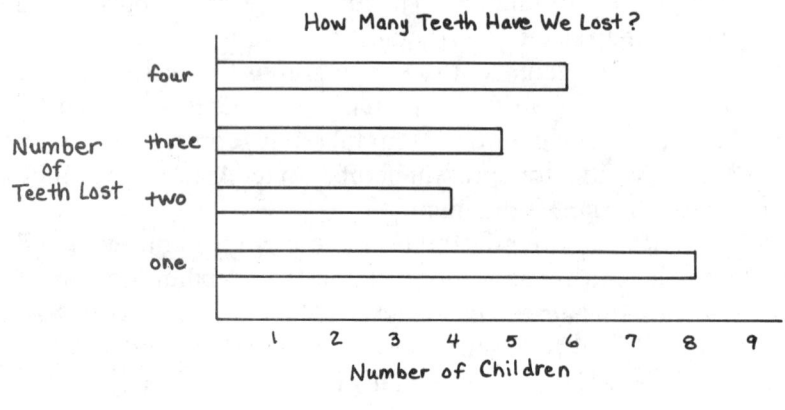

Fig. 2

Multiple or double bar graphs are used to compare discrete stratified data (i.e., data collected from particular groups). For example, when asking children to vote for their favorite pets, colors, or favorite games to play, organize the results according to boys' responses and girls' responses. Double bar graphs are presented in Activities 9, 11–13, and 19–21. See figure 3 for a sample of this type of graph, based on the data in figure 1.

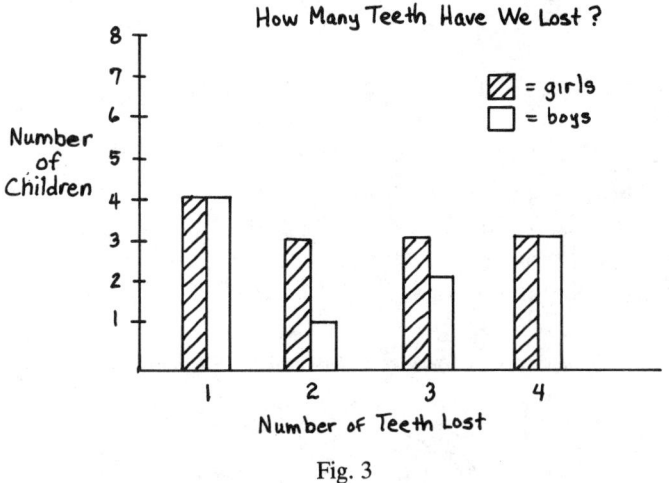

Fig. 3

Line graphs. A line graph (also called a broken-line graph) is used to compare continuous data. Points are plotted within perpendicular axes to represent change over a period of time or any linear functional relationship. The axes, which are labeled, intersect at a common point, usually zero. The units of division on each axis are equally spaced. The graphed points are connected by straight or broken lines.

Children can keep a record over a period of time of their own height or weight, of the daily average temperature, of the height of a plant, and so on. Ideas for discussing the uses of line graphs are in Activities 22–25. See figure 4 for an example.

Multiple line graphs are used to compare two or more sets of continuous data—for example, to compare the heights or weights of two children over a period of time (e.g., four months or one year), or the heights of two (or more) plants over a period of time (e.g., one to two months after planting seeds). Ideas for presenting multiple line graphs are in Activities 24 and 25. See figure 5 for an example.

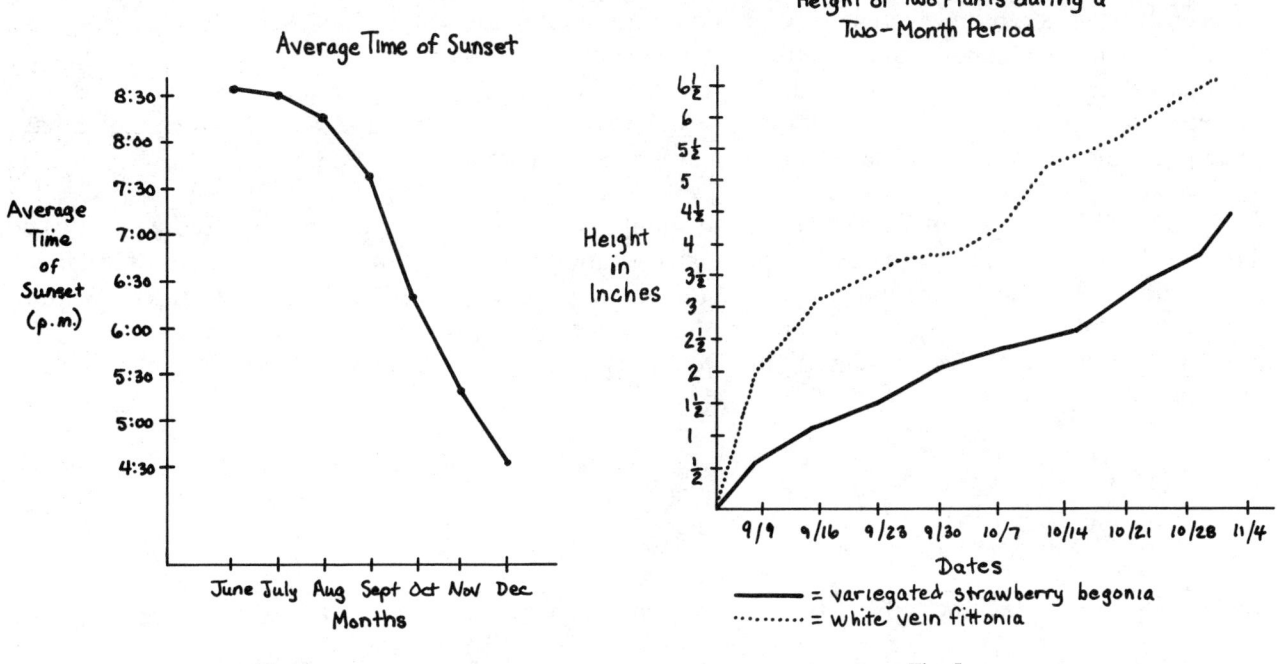

Fig. 4

Fig. 5

Circle graphs. The area of the circle graph (also called a pie graph, pie chart, pie diagram, or area graph) is divided into sectors, depicted by lines emanating from the center of the circle. "Each [sector] represents a proportionate part of the whole" (Arkin and Colton 1940, p. 131). The circle graph is used when data are to be compared to the whole and different parts of the whole.

Presenting circle graphs to children is appropriate after they have an understanding of fractions. Successful construction of a circle graph is dependent on children's understanding of proportion and their ability to use a compass and protractor. Data for such topics as budget and population characteristics are sometimes displayed in circle graphs. Of all the traditional graph types, the circle graph is the most difficult to construct.

Although statisticians and mathematics educators indicate that there should be less emphasis on circle graphs in the curriculum today (e.g., James Landwehr, personal communication, 2 June 1987; Albert Shulte, personal communication, 7 April 1988), there is still a proliferation of them in the media; consequently, this graph form should remain in the curriculum. Ideas for discussing circle graphs are in Activities 16–18. See figure 6 for an example.

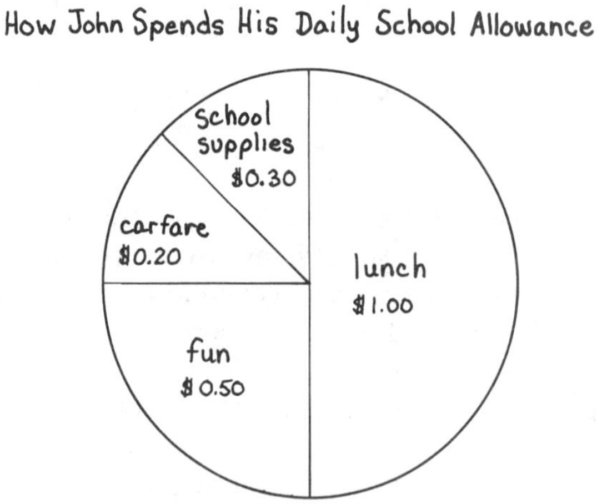

How John Spends His Daily School Allowance

Fig. 6

Some New Techniques

Recently, suggestions have been made for incorporating the use of line plots, stem-and-leaf plots, and box plots in the elementary and middle school curriculum (Corwin and Friel 1988; Landwehr and Watkins 1986; Silverman 1988). The development of the stem-and-leaf plot and the box plot has been attributed to John Tukey (1977). It has been noted that these new techniques for displaying data have certain advantages over the traditional graphical forms (Landwehr and Watkins 1986).

Line plots. Line plots look like primitive bar graphs, where numerical data are plotted as *x*'s placed above numbers on a number line. A line plot "gives a graphical picture of the relative sizes of numbers, and it helps you to make sure that you aren't missing important information" (Landwehr and Watkins 1986, p. 5). Unlike a bar graph, in which data may be lost in the grouping, none of the data gets lost in a line plot. Usually the number of items plotted does not exceed twenty-five. See figure 7 for an example of a line plot.

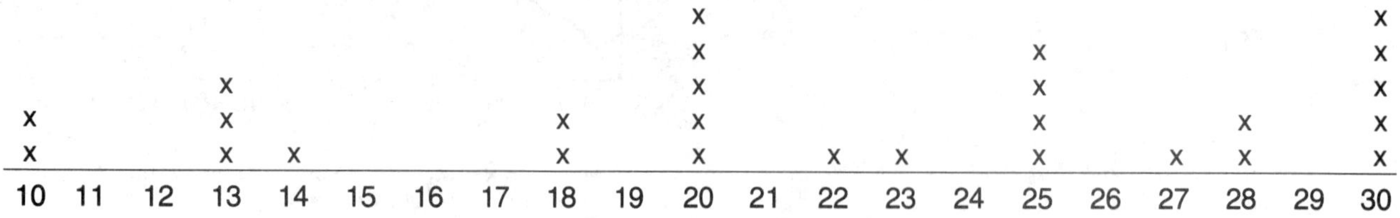

Fig. 7. Estimates for the number of beans in a handful, depicted on a line plot

Stem-and-leaf plots. These plots are characterized by a "separation" of the digits in numerical data. For example, in a simple stem-and-leaf plot, appropriate for fifth graders, the tens digits are listed in one column and the ones digits are listed in a row next to the respective tens digit (see fig. 8). When rotated ninety degrees counterclockwise, the stem-and-leaf plot resembles a bar graph. This plot "is often better than the bar graph . . . because it is easier to construct and all the original data values are displayed" (Landwehr and Watkins 1986, p. 9).

Back-to-back stem-and-leaf plots are more complicated for students to read. The ones digits of another set of data are attached to the left-hand side of a stem-and-leaf plot (see fig. 9).

```
 Tens | Ones
   1  | 00333488
   2  | 00000235555788
   3  | 00000

        1 | 0   means 10
```

Fig. 8. Estimates for the number of beans in a handful, depicted in a stem-and-leaf plot

```
         Actual Count              Estimates
              Ones | Tens | Ones
         9999998888 |  1   | 00333488
   99553222211111110 |  2   | 00000235555788
                    |  3   | 00000
                              8 | 1 | 0   means
                              18 actual count
                              10 estimate
```

Fig. 9. Actual counts and estimates of beans in a handful, depicted in a back-to-back stem-and-leaf plot

Box plots. Box plots (also referred to as box-and-whisker plots) use five summary numbers (i.e., the lower extreme, the lower quartile, the median, the upper quartile, and the upper extreme) and are helpful when analyzing large quantities of data (i.e., more than 100 pieces of data). Although this type of display may be more difficult to construct, it has been used effectively with middle school students.

When using a box plot, "we can no longer spot clusters and gaps, nor can we identify the shape of the distribution as clearly as with line plots or stem-and-leaf plots. However, we are able to focus on the relative positions of different sets of data and thereby compare them more easily" (Landwehr and Watkins 1986, p. 73). See figure 10 for example of a box plot.

These three new plotting techniques are presented in Activity 21. Teachers are encouraged to use these ideas in other data-gathering and data-analyzing activities.

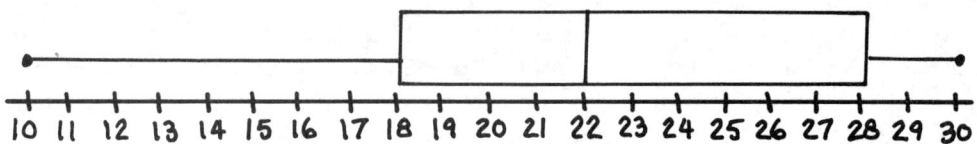

Fig. 10. Estimates for the number of beans in a handful, depicted in a box plot

LEVELS OF GRAPH COMPREHENSION

Although literal reading of data presented in graphical form is an important component of graph-reading ability, the maximum potential of the graph is realized when the reader is capable of interpreting and generalizing from the data (Kirk, Eggen, and Kauchak 1980). Supporting the reading comprehension literature, the results of a recent research study suggest that there are three distinct levels of graph comprehension (Curcio 1987). Regardless of the graph form used, the three levels of graph comprehension are reading the data, reading between the data, and reading beyond the data.

Reading the Data

This level of comprehension requires a literal reading of the graph. The reader simply "lifts" the facts explicitly stated in the graph, or the information found in the graph title and axis labels, directly from the graph. There is no interpretation at this level. Reading that requires this type of comprehension is a very low level cognitive task. Questions 1 and 2 for all the supplemental graph-reading activities in Appendix 9 are examples of reading-the-data questions.

Reading between the Data

This level of comprehension includes the interpretation and integration of the data in the graph. It requires the ability to compare quantities (e.g., greater than, tallest, smallest) and the use of other mathematical concepts and skills (e.g., addition, subtraction, multiplication, division) that allow the reader to combine and integrate data and identify the mathematical relationships expressed in the graph. This is the level of comprehension most often assessed on standardized tests.

Reading between the data requires "at least one step of logical or pragmatic inferring necessary to get from the question to the response *and* both question and response are derived from the text" (Pearson and Johnson 1978, p. 161). Questions 3 and 4 for all the graphs in Appendix 9 are examples of reading-between-the-data questions.

Reading beyond the Data

This level of comprehension requires the reader to predict or infer from the data by tapping existing schemata (i.e., background knowledge, knowledge in memory) for information that is neither explicitly nor implicitly stated in the graph. Whereas reading between the data might require that the reader make an inference based on the data presented in the graph, reading beyond the data requires that the inference be made on the basis of a "data base" in the reader's head, not in the graph. Questions 5 and 6 for all the graphs in Appendix 9 are examples of reading-beyond-the-data questions.

A section entitled "Questions for Discussion" is included in each of the classroom activities. The questions intended for reading the data (RD), reading between the data (RBW), and reading beyond the data (RBY) are identified as such. However, the level of comprehension for open-ended questions cannot be determined until a response is given (Pearson and Johnson 1978). Different children may interpret the questions differently.

In addition to the sample questions provided in each activity, teachers are encouraged to formulate other questions that require reading between and beyond the data. Also, at the end of each question section, a suggestion is made for students to think of questions that relate to the graph being analyzed.

The graphs developed by students are not to be treated as "static displays, an end in themselves," but rather the graphs should lead to further questions to enhance the development of critical thinking (Russell 1988, p. 4). It is hoped that some reading-beyond-the-data questions may lead to further investigation and inquiry (see Activities 1, 2, 4, 7, 11, 13, 14, and 18).

COLLECTING, ORGANIZING, AND ANALYZING DATA

Collecting Data

It is never too early to involve children in actively collecting data to construct their own graphs. The activities should be interesting and meaningful and should reflect the experiences of children at their particular grade, age, and ability levels. Topics suggested for different grade levels can be found in Appendix 1. However, it is important for teachers to determine whether the topics are relevant for the children in their classes.

In the early elementary grades, one way to involve children in data collection is to ask them a question about a particular interest. For example, such questions as "What is your favorite color?" "What is your favorite pet?" will help to elicit personal, relevant information. Whether children respond orally, by raising their hands, by attaching a piece of paper on the chalkboard, or by casting a ballot, once they begin to answer the question, the problem of how to organize everyone's responses becomes evident. The amount of data collected and the data values should be meaningful and manageable for the children.

In the middle grades, the class survey or poll is a way to involve children in data collection (Newman and Turkel 1985). After discussing with students a topic that interests them, allow them to formulate survey questions. Before including the questions in their survey, students should discuss them and come to an agreement about phrasing, relevance, importance, and clarity. The amount of data collected and the data values should be meaningful and manageable for the students. Prior to conducting the survey they should discuss how the information they collect will be organized and analyzed. By conducting a survey, students can appreciate the life of a pollster!

Another source of data for middle school students is the library. Reference yearbooks, newspapers, almanacs, and records and facts books can provide a rich source of data. Prior to referring to such sources, students should formulate questions of interest for which they need to seek answers. At this point, the data values involved may be very large (i.e., in the millions). It is important that students have an appreciation and understanding of the large numbers they will encounter.

Often the results of national surveys are reported in graphical form in the newspapers. If the topic of a national survey is appealing to students, involve them in conducting a similar survey to collect data locally and compare the results to the newspaper graph (see Activities 11, 13, 14, and 18).

All the activities in this book are meant to be interesting and relevant, and they involve data values that are meaningful and manageable for students. Teachers should make adjustments so that the activities meet the needs of the individual children.

Organizing Data

Once the data are collected, children should be aware of the need to organize the data in a meaningful way so that they can be interpreted and analyzed. In the very early grades the number of data should be countable within determined categories.

In the upper elementary and middle grades, children may need to learn tallying techniques. Teachers should not assume that children know how to tally efficiently. Although tallying is not difficult, children should be encouraged to discuss how they might go about tallying the data. Some children may have ideas to share with others.

Organizing the data in a chart or table is a very helpful way of getting a "picture" of the data. The occurrence of certain numbers or patterns will help students decide how to set up a graph for the data. Getting a "feel" for the data is necessary in determining which graph form is appropriate for displaying them. It is important to note that tables and charts are often presented in newspapers without a visual display. However, presenting the data in a graph can often reveal patterns and trends not readily visible in the chart or table.

Analyzing Data

Once the data are organized, students should determine whether they are presented adequately for examining patterns, relationships, or trends. Explaining what the data mean explicitly is necessary before examining the implicit relationships between and among the data. Information presented in a chart or table may need to be converted to a graph to read between and beyond the data. Can any predictions be made? Can any "What if?" questions be asked? Are there any questions that arise from the data collected?

In each of the activities in this book, "Questions for Discussion" are meant to guide students in reading between and beyond the data. If the students are interested, they may want to conduct another survey and construct another graph. They should be encouraged to pursue related projects and report their findings to the class.

CONSTRUCTING, INTERPRETING, AND WRITING ABOUT GRAPHS

Constructing Graphs

Graphing ideas develop naturally from activities that involve sorting and classifying (Baratta-Lorton 1976; Bruni and Silverman 1975). As early as kindergarten, children should be involved in collecting data about themselves (e.g., eye color, hair color, favorite color). By using a floor grid (see Appendix 2), children can create a people graph by lining up according to their appropriate characteristic. For example, all the children with the same hair color or eye color can line up on the floor grid behind each other (see Activities 1 and 2).

People graphs lead naturally to block graphs, where children can color, for instance, a pair of eyes drawn on an index card. The index card can be attached to a block (i.e., same-size blocks from the block corner) and children can pile similar data blocks on top of each other (see Activity 2).

People graphs also lead naturally to object graphs, where realia can be used to represent the appropriate data (Choate and Okey 1981). For example, toy cars, toy buses, and doll shoes can be used to represent the modes of transportation to school. These toys can be attached to a grid made from oaktag (see Appendix 2), which can then be displayed on a bulletin board or on the chalkboard (see Activity 4).

On the basis of these primitive graphs (i.e., people graphs, block graphs, and object graphs), picture graphs can be developed. Using uniform ideographs (e.g., pictures of eyes organized by color), children can identify the one-to-one correspondence represented. Eventually, by the third grade, children should be exposed to pictorial representations of simple many-to-one correspondences, for example, where two pairs of eyes are represented by one ideograph (see Activities 4–7 and 9–11).

Life-sized bar graphs can be created as early as the first grade. Children working in pairs can outline their bodies on poster paper or newsprint, cut out their shapes, and create a height bar graph. To avoid confusing children with too much data at one time, no more than four height bars should be displayed on the same life-sized graph. Following this activity (or in lieu of it), children's heights can be represented by lengths of adding-machine tape. Again, no more than four bars should be displayed on the same graph (see Activity 3).

Converting a picture graph to a bar graph is a natural way to help children move from a semiconcrete representation of data to a form that is more abstract. When the teacher recognizes that the children are ready, the picture groupings of one-to-one correspondences on a transparency, chalkboard, or piece of poster paper can be outlined by bars. Then a labeled, numerical, vertical or horizontal axis should be inserted and the pictures removed (see Activities 5–7, 9, and 11).

In the middle grades, children should be involved in discussing the nature of the data collected (i.e., continuous vs. noncontinuous data) and from that, they should determine the appropriate type of graph to display the data. If traditional graph forms are used (i.e., picture, bar, line, and circle graphs), the children should determine the variables, the labels for the axes, and the title of the graph and then proceed to plot the data.

Students should compare data displayed in various graph forms. They should identify when the data are seemingly distorted or when they seem to be represented fairly. Having access to computers can facilitate graph construction so that students can focus on comparing and analyzing the graphs rather than on constructing them (see "Using the Computer as a Tool for Graphing and Analyzing Data," p. 10).

Also, it is sometimes desirable to display in a circle graph data originally presented in a bar graph. Students should examine the similarities and differences of alternative ways of presenting the data (see Activity 18).

If nontraditional plots are used (i.e., line, stem-and-leaf, and box plots), students must determine which plot would be most appropriate. If they decide to construct a line plot, they must be able to identify the extremes, proceed to construct a number line, and then place an x for each piece of data in the proper place on the number line. If a stem-and-leaf plot is to be employed, the children must be able to set up the place value rows properly and list the data appropriately. If a box plot is to be employed, they should be able to identify the median, the lower and upper quartiles, and the lower and upper extremes (see Activity 21).

Interpreting Graphs

As young children experience the physical creation of a graph, they should be involved in interpreting it. Questions that reflect reading the data, reading between the data, and reading beyond the data provide a basis for interpreting and discussing graphs.

Discussion about graphs in the early grades should revolve around the language arts, that is, listening, speaking, reading, and writing. While the class listens, individual children should be encouraged to talk about the graphs they created, and the teacher (or an adult aide) should record the children's comments. Depending on children's reading skills, they can read the comments individually or in unison.

These ideas should continue through the middle grades. Once graphs are constructed, children should be involved in answering questions "that lead to prediction, interpretations, and to additional questions" (Russell 1988, p. 9). Working in groups of four or five students, they should talk about the graphs they create, exchange them, and constructively criticize and question their peers.

Writing about Graphs

Activities that provide children with the opportunity to interpret graphs and plots should include teacher- and student-formulated questions reflecting different levels of comprehension. On the basis of these questions, children should be encouraged to write a paragraph about the graph or plot. Writing about graphs and plots allows children to clarify their thinking and communicate their interpretation with others. Both the graphs and the written descriptions should be shared among the students. They should be encouraged to question and criticize each other constructively. Writing about graphs is a summary feature for each activity in this book.

CLASSROOM ACTIVITIES

Overview

This section contains twenty-five field-tested classroom activities. The activities are meant to be developmental and are organized according to grade level. The levels, indicated in the Table of Contents, are meant to be a guide for planning instruction. Depending on children's ability, teachers may have to adjust or modify the activities to meet their needs.

Although the activities contain many components, they are not intended to be presented to all children, nor are they intended to be accomplished during one lesson. The activities present ideas for developing graphing concepts for children with varying degrees of background knowledge and experiences. The amount of time spent on each component should vary according to the children's ability.

For each activity there are sections that identify the topic, the graph form to be presented, a graph title, objectives, vocabulary, suggested materials for presenting each graph form, procedures for developing the ideas, using the computer as a tool (beginning with Activity 8), questions for discussion, and ideas for summarizing the activity. These sections are repeated in each activity, sometimes verbatim, so that the activities can be used independently.

Many of the topics were determined on the basis of a survey of twenty-four elementary and middle school teachers from the New York metropolitan area, conducted during 1987. Each teacher had over five years of teaching experience at multiple grade levels. Appropriate grade levels for presenting the graph topics were agreed on by at least 75 percent of the teachers (see Appendix 1 for an extensive list of topics). Supplemental graph-reading activities are provided in Appendix 9. These are not meant to replace data collection and analysis but rather to supplement the hands-on experiences provided in the classroom activities.

Most of the graph forms presented are traditional forms (i.e., picture, bar, line, and circle graphs). The graph forms presented in Activities 1–7 are not formal graphs; they are meant to be informal pregraphing activities. A more formal treatment of graphing begins in Activity 8, in which using the computer as a tool is presented as an option for teachers. Some activities suggest the presentation of several graph forms so that children can observe and discuss similarities and differences between and among the different graph forms. Ideas for presenting multiple graph forms appear in Activities 2–7, 9–12, 18, 21, 24, and 25. Teachers should decide which graph forms are appropriate as well as when connections between and among multiple forms should be made.

One activity presents ideas for applying some of the new plotting techniques. Activity 21 focuses on the line plot, stem-and-leaf plot, back-to-back stem-and-leaf plot, box plot, and multiple-box plot. This activity is not meant to be an introduction to these new techniques but rather to provide a sample application. Teachers are encouraged to apply these to some of the other activities.

In the activity plans, the graph title is usually presented in question form. It is the question that often helps to stimulate the activity. After creating a graph, children should identify the graph title. Whether the title is displayed on the chalkboard, the overhead projector, poster paper, or their own graphs, allow children to select the phrasing of the title.

The objectives present the goals of the different components of the activities in terms of student outcomes. The procedure, questions for discussion, and summary are built around these objectives.

Developing vocabulary should be a part of every mathematics lesson. Graphing activities should be no exception. Depending on children's ability, the vocabulary words may or may not be familiar to them. New words should be discussed and familiar words should be reviewed. The words should be used in the writing and reading tasks suggested in the summary.

The materials needed to conduct the activity are listed. However, teachers may feel that they would like to add or delete certain parts of the activity. Also, there may be other materials that have been inadvertently omitted. These should be checked carefully before beginning the activities with the children. Graph paper with different sizes of boxes appropriate for different grade levels can be found in Appendixes 4–7. A 24-section circle graph outline is in Appendix 8.

The "Procedure" presents some opening questions for discussion and an outline concerning how the activity is to proceed. Teachers should feel free to deviate from this and experiment with their own ideas.

"Questions for Discussion" list questions that reflect different levels of comprehension. Abbreviations are used after the questions that apply: Reading the Data (RD), Reading between the Data (RBW), and Reading beyond the Data (RBY). It is important to note that although questions may be intended for a particular level, the actual level of comprehension cannot really be identified until the student's response is analyzed (Pearson and Johnson 1978). Other graphing activities that contain questions aimed at these levels of comprehension can be found in Appendix 9.

As a means to enhance children's ability to communicate with mathematics, writing and reading skills should be developed and reinforced in the mathematics lesson. Writing about the activity and the graph, reading it aloud or sharing it among students, and allowing students to compare the paragraph with the graph affords students the opportunity to clarify their thinking and communicate their interpretations about the graphs with others. This should be a part of every graphing activity.

Using the Computer as a Tool for Graphing and Analyzing Data

With the advent of microcomputer technology, students must learn how to access and display information. Beginning with Activity 8, "Using the Computer as a Tool" provides suggestions for graphing and analyzing data collected by children. Depending on the students' graphing experiences, the teacher may want to develop ideas using the computer rather than having children construct graphs by hand. However, to understand and appreciate fully the power of the computer to graph data, students must have adequate experience using manipulatives and constructing their own graphs (Mathis 1988a).

At the time the teacher believes that students are ready to use the computer, they may or may not be able to recognize the characteristics of a set of data (i.e., discrete vs. continuous). In either event, students should display the data in different graph forms to analyze similarities and differences. Graphs that seemingly distort the data and those that represent the data "fairly" should be discussed. Whether the intent is to focus on the meaning of the data as it is represented in particular graph forms or to compare the advantages and disadvantages of different displays, using the computer in this way will free students to spend more time interpreting and analyzing graphs rather than constructing them.

To identify graphing software for use with elementary and middle school students, teachers must keep abreast of the development of new products. Software should be constantly and consistently reviewed and evaluated to determine whether it is appropriate for use as a tool for graphing data collected by children.

Stone (1988, p. 16) has recommended seven factors to consider when reviewing and selecting graphing software tools. A graphing utility should do the following:

1. Provide a choice among a variety of graphing formats and enable alternate representations of a given data set
2. Provide for simple, straightforward data entry and editing information
3. Maximize student control of labeling, range and number of entries, scaling, and format
4. Make relevant on-screen help readily available
5. Produce clear, easily understandable printouts
6. Allow disk storage of both data set and graph
7. Execute graphs accurately

Although a variety of graphing software is commercially available and new products will most likely become available in the future, there are currently a limited number of good graphing utilities available for use with elementary and middle school students. In particular, table 1 identifies *MECC Graph* (Edusystems, Inc. and MECC 1985), *Easy Graph II* (Lewis et al. 1987), and *Exploring Tables and Graphs* (Bannasch 1984) as having some of the features that are deemed appropriate for instructional use (Mathis 1988a, 1988b). For a more detailed review of these utilities, see Mathis (1988a). The classroom activities for which these graphing utilities are appropriate are also indicated in table 1.

Software developed in conjunction with the Quantitative Literacy Project (Stanek and Johnson n.d.) allows students to construct stem-and-leaf and box plots. Although it does not have all the desirable features recognized by Stone (1988), it would be appropriate for limited use in Activity 21.

The software identified here is offered only as a sample of what is available. Depending on the types of computers in their schools, teachers should continue to examine and select appropriate graphing software on the basis of Stone's (1988) criteria.

TABLE 1
Features of Selected Pieces of Software That Support Classroom Activities

Title/(Computer)	Grade Level	Types of Graphs	1[a]	2	3	4	5	6	7	Files Saved	Activities
Easy Graph II (Apple II, 64K)	K–4	Picture, Bar, Circle	+[b]	+	–	0	+	0	+	Graph	8–12 (partially), 15–18 (partially)
Exploring Tables & Graphs I	3–4	Picture, Bar (vertical & horizontal), Circle, Line (II only), Tables	+	0	–	0	0	0	+	Data	8–12, 15
Exploring Tables & Graphs II (Apple II, 48K)	5–6		+	0	–	0	0	0	+	Data	8–12, 15, 17, 19, 20, 24
MECC Graph (Apple II, 64K)	3–12	Bar, Circle, Line	+	+	+	0	+	0	+	Data	8, 9, 11 (partially), 12, 15–20, 22–24

[a] 1 = Provide a choice among a variety of graphing formats and enable alternate representation of a given data set.
2 = Provide for simple, straightforward data entry and editing of information.
3 = Maximize student control of labeling, range and number of entries, scaling, and format.
4 = Make relevant on-screen help readily available.
5 = Produce clear, easily understandable printouts.
6 = Allow disk storage of both data set and graph.
7 = Execute graphs accurately (Stone 1988).
[b] + = Met
0 = Partial
– = Not Met

Excerpts from Mathis (1988b) reprinted with permission from *The Computing Teacher*, November 1988, p. 10. Published by the International Council for Computers in Education.

ACTIVITY 1

Topic: Hair color

Graph Form: People graph

Graph Title: What Color Hair Do We Have?

Objectives
1. To identify hair color
2. To collect, organize, and interpret data
3. To create a people graph
4. To categorize oneself in a people graph according to hair color
5. To answer comprehension questions based on the people graph
6. To write a story based on the graph

Vocabulary: Graph, people graph, hair colors (review if necessary), more, most, fewest

Materials: Floor grid (see Appendix 2); 5–7 pieces of 8½″ × 11″ paper to use as floor grid labels; paint, crayons, or colored markers to write hair colors on floor grid labels; writing tablet

Procedure
Ask, "What color hair do most of the children in the class have? How can we be sure? What are the different hair colors?" Elicit from the children: brown, black, blond, red, and so on. Use appropriate colors of paint or crayons to write the colors on pieces of paper to fit at the base of the floor grid.

Instruct the children with brown hair (for example) to line up behind the brown label, those with black hair to line up behind the black label, and so on. Have other pupils count the number of children in each row, identifying the number with each hair color.

Questions for Discussion
1. "How many children have brown hair? All the children with brown hair raise your right hand." Allow the rest of the children to count those with brown hair and record the number on the chalkboard. Do the same for other colors. [RD]

2. "What title can we give our graph?" Write the title on poster paper, the chalkboard, or on the writing tablet for all the children to see. [RBW]

3. "For which color are there the most children? How do you know this?" [RBW]

4. "For which color are there the fewest children? How do you know this?" [RBW]

5. "Do you think there are more children in all the kindergartens (or first-grade classes) with blond hair or brown hair? Can you answer this question from the graph? Why? Why not? How can we find out the answer to this question?" [RBY] This could lead to collecting information from another kindergarten or first-grade class.

6. "What question can you ask about our graph?"

Writing and Reading
Once the children have responded to the questions, ask them to tell a story about the experience. Write the story as the children dictate it, and then help them read it together.

ACTIVITY 2

Topic: Eye color

Graph Form: a. People graph
 b. Block graph (with pictures of eyes colored appropriately)
 c. Picture graph

Graph Title: What Color Eyes Do We Have?

Objectives
1. To identify eye color
2. To collect, organize, and interpret data
3. To create a people graph, block graph, or picture graph
4. To categorize oneself in the graphs according to eye color
5. To answer comprehension questions
6. To write a story from the graph

Vocabulary: Graph, people graph, block graph, picture graph, eye colors (review if necessary), more, most, fewest

Materials
 a. Floor grid (see Appendix 2); ready-made floor grid labels of eye colors (printed in appropriate colors on white paper, 8½" × 11"; construction paper for children to represent their eye colors; writing tablet; marker
 b. Blocks (same size, one for each child); index cards cut to fit the side of the blocks (for children to draw and color their eyes); crayons (to match eye colors); tape (to attach cards to blocks); space on a table or desk to construct a block graph for all the children to see; ready-made block graph labels with eye colors; writing tablet; marker
 c. Construction paper (3" × 5", one piece for each child to represent his or her appropriate eye color); tape (to attach each piece of colored construction paper to a picture graph on the chalkboard); writing tablet; marker

Procedure
Ask, "What color eyes do most of the children in our class have? How can we be sure? What are the different eye colors? [e.g., brown, blue, green, hazel, gray] What color eyes do you have?" For the people graph, distribute pieces of construction paper to match individual eye colors, and have the children print their names on their sheets. For the block graph, distribute index cards and crayons for children to draw and color their eyes. Attach the cards to the blocks.

 a. *People graph:* Place the ready-made floor grid labels at the base of the grid (see fig. 2.1). Instruct the children with black eyes to line up behind the black-eyes label, children with blue eyes to line up behind

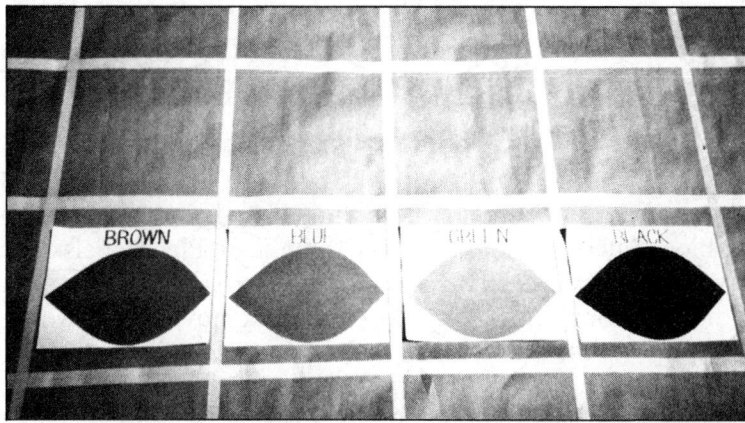

Fig. 2.1

the blue-eyes label, and so on (see fig. 2.2). Have the children with black eyes raise their pieces of black construction paper. Allow the other children to count those with black eyes and record the number on the chalkboard. Do the same for the other eye colors.

After asking some of the questions listed below, have the children step off the floor grid, leaving their pieces of construction paper in the places where they were standing (see fig. 2.3). Continue with the questioning. Allow the children to sit around the graph, making up a story about the activity (see fig. 2.4). The teacher should write the story that the children dictate on the chalkboard or on poster paper (see fig. 2.5).

Fig. 2.2

Fig. 2.3

Fig. 2.4

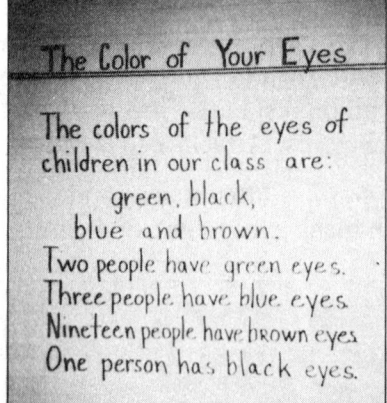

Fig. 2.5

b. *Block graph:* Have the children draw a pair of eyes on an index card and color them to match the color of their eyes. Attach each index card to a block (all blocks should be the same size). Place labels on a table to identify the various eye colors, and have the children stack their blocks in the appropriate pile (see fig. 2.6). After discussing answers to some of the questions below, have the children remove the drawings of their eyes from the blocks (see fig. 2.7); continue with the questioning, at the same time eliciting a story about the activity from the children (see fig. 2.8).

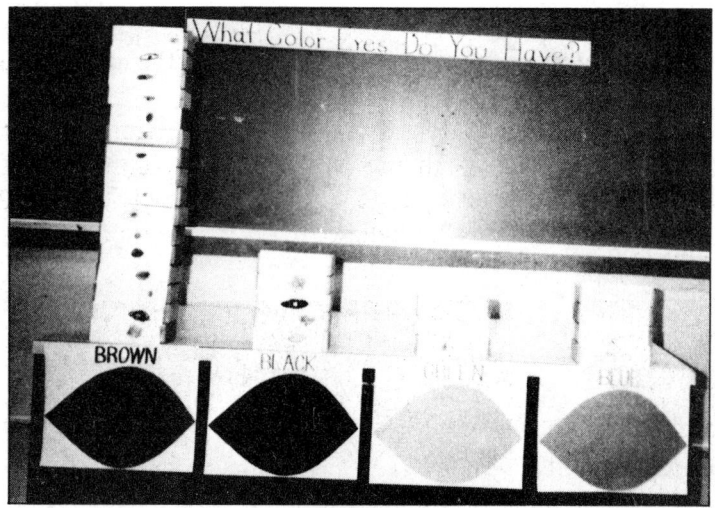

Fig. 2.6

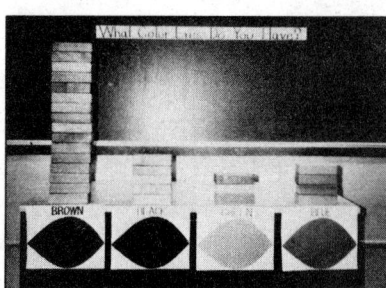

Fig. 2.7

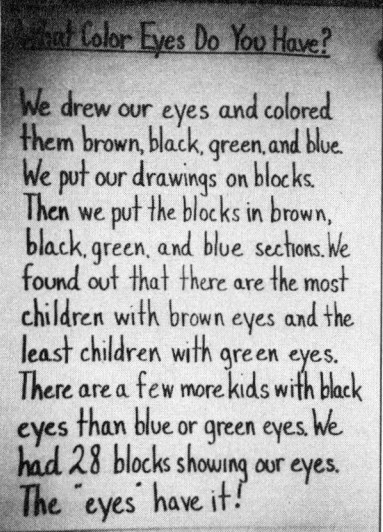

Fig. 2.8

c. *Picture graph:* Allow each child to select a piece of 3″ × 5″ construction paper to match the color of his or her eyes. On the chalkboard, have the children attach their pieces of construction paper in the appropriate row. Labels of different eye colors should identify the colors represented in each row.

Questions for Discussion

1. "What title can we give our graph?" Write the title on the chalkboard or on paper to post near the graph so all the children can see it. [RBW]

2. "How many children have black eyes? Brown eyes? Blue eyes?" [RD]

3. "What color eyes do most of the children in our class have? How can we be sure?" [RBW]

4. "What color eyes do the fewest number of children in our class have? How do you know this?" [RBW]

5. "What color hair do most people with brown eyes have?" [RBY]

6. "What color hair do most people with blue eyes have?" [RBY]

7. "Do you think there are more children in our school [or all the kindergartens or first-grade classes] with brown eyes or blue eyes? Why do you think this? Can you answer this question by just looking at the graph? What could we do?" [RBY] This could lead to collecting information from another kindergarten or first-grade class.

8. "What questions can you ask about this graph?"

Writing and Reading

Once the questions have been discussed, ask the children to tell a story about the activity. Write their story on the chalkboard or on poster paper (see figs. 2.5 and 2.8) and encourage the children to read their story aloud.

ACTIVITY 3

Topic: Height

Graph Form: Life-sized bar graph
      a. Using outlines of children's bodies
      b. Using adding-machine tape

Graph Title: How Tall Are We?

Objectives
1. To compare heights by standing next to each other
2. To construct life-sized models to compare heights
3. To use adding-machine tape to represent and compare heights
4. To collect, organize, and interpret data
5. To answer comprehension questions
6. To write stories based on the life-sized graphs

Vocabulary: Graph, height, tall, taller, tallest, short, shorter, shortest

Materials
   a. Large pieces of drawing paper (or wrapping paper) to outline the children's bodies; markers; scissors; tape, tacks, or paper clips (to display outlines)
   b. Adding-machine tape to represent the measure of the children's heights; markers; scissors; tape, tacks, or paper clips (to display adding-machine tape)

Procedure
Ask, "Who is the tallest student in our class? How can we be sure?" Have students compare heights by standing next to each other in pairs, threes, and fours.

   a. *Outline of children's bodies:* Have children lie on large pieces of drawing paper, newspaper (taped together), or wrapping paper while their partners draw outlines of their bodies. Have the children cut out their body outlines and write their names on them. Use three or four outlines at a time to display on a wall or bulletin board as a life-sized graph. Be sure the base of each outline is at the same level for proper comparison. Change the outlines periodically, reviewing the questions below when the changes are made.

   b. Have pairs of students use adding-machine tape to represent each child's height, from foot to head. Help the children make sure that the tape does not sag, thus outlining the child's body, but rather is held taut and perpendicular to the floor. The piece of adding-machine tape can be folded and placed under the child's foot so that the tape can be held taut. Have each child cut the proper length to represent his or her partner's height, and have the children write their names on their tapes. Use three or four lengths of tape at a time to display on a wall or bulletin board as a life-sized graph, being sure that the base of each tape is at the same level for proper comparison. Change the tapes periodically, reviewing the comprehension questions when the changes are made.

Questions for Discussion
1. "What would be a good title for this graph?" Write the title above the graph, being sure to include the month and year. [RBW]
2. "What are the names of the children whose heights are represented in the graph?" [RD]
3. "How tall is [insert a name]?" [RD]
4. "Who is the tallest of the students on the graph?" [RBW]
5. "Who is the shortest?" [RBW]
6. "Who do you think is the oldest? Why? Can this be answered directly from the graph?" [RBY]

7. "Who do you think has the smallest shoe size? Why? Can this be answered directly from the graph?" [RBY]
8. "Who do you think weighs the least? Why?" [RBY]
9. "What are some questions you can ask about this graph?"

Writing and Reading

After the questions are discussed, ask the children to tell a story about the activities. Have them write their own stories, or use the children's words to write the story on the board or poster paper so that all the children can see it. Encourage them to read the story together.

ACTIVITY 4

Topic: Travel

Graph Form: a. Object graph
b. Object/person picture graph
c. General picture graph

Graph Title: How Do We Get to School?

Objectives
1. To identify ways of traveling to school
2. To collect, organize, and interpret data
3. To use objects to represent the mode of travel and to recognize a one-to-one correspondence between the children and the objects
4. To recognize the replacement of a picture (either of an object or a person) to represent how one travels to school
5. To recognize the replacement of a general object (e.g., a cube) for a specific object in a graph
6. To construct an object graph and a picture graph
7. To answer comprehension questions
8. To write a story based on the graph

Vocabulary: Graph, object graph, picture graph, most

Materials

a. Ready-made chart for object graph (see Appendix 2); model or toy shoes (to represent walkers), buses, cars, and trains; pictures of ways children travel to school (Appendix 3); paper clips (to attach pictures to ready-made chart); tape to attach objects; Unifix cubes to represent a generalization of modes of travel; lined paper for writing graph stories

b. Ready-made chart for picture graph (see Appendix 2); pictures of children or drawings to represent modes of travel to school; paper clips (to attach pictures to ready-made chart); lined paper for writing graph stories

c. Ready-made chart for picture graph (see Appendix 2); uniform geometric shapes cut from construction paper (i.e., same size, shape, and color) to generalize modes of transportation; paper clips (to attach paper to ready-made chart); lined paper for writing graph stories

Procedure

Ask, "What are the different ways we can travel to school?" List the children's responses on the chalkboard or poster paper.

a. *Object graph:* Ask individual children the way they travel to school. Have each child pick up a representative model or toy and attach it with tape to the ready-made chart (see fig. 4.1). After discussing the object graph, have the children replace the objects with Unifix cubes to represent the modes of transportation.

b. *Picture graphs:* Ask the children to draw pictures to represent the way they travel to school or to bring in pictures of themselves. Have them attach their pictures to the ready-made chart to indicate which mode of transportation they use to get to school (see fig. 4.2).

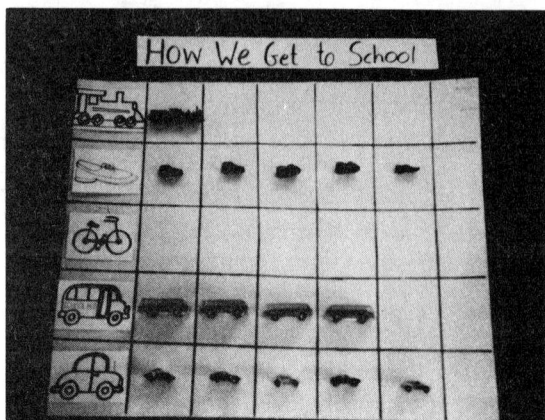

Fig. 4.1

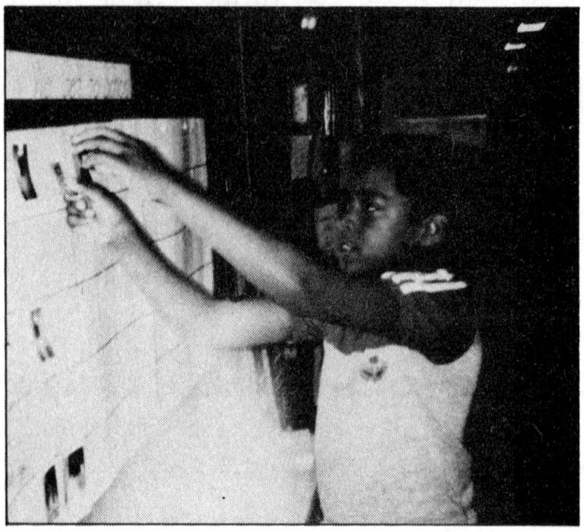

Fig. 4.2

c. *General picture graphs:* After discussing the picture graphs, have the children replace their pictures with uniform pieces of colored paper (any geometric shape).

Questions for Discussion

1. "What is a good title for this graph?" [RBW]
2. "How many children walk to school?" [RD]
3. "How many children ride a bike to school?" [RD]
4. "How many children ride the bus to school?" [RD]
5. "How many children are driven to school in a car?" [RD]
6. "Which is the most popular way for children to travel to school? How do you know? How can you be sure?" [RBW]
7. "Do you think this graph would change in any way if it were raining or snowing? How do you think it would change?" [RBY]
8. "About how many children live close to school? How do we know this?" [RBY]
9. "What are some questions you can ask about this graph?"

Writing and Reading

After the children have discussed the answers to the questions, ask them to tell a story about the activity. Using the children's words, write the story on the board or poster paper (see fig. 4.3), and help the children read the story aloud together. By the third grade, children should be able to write their own stories (see fig. 4.4).

How We Get to School

In our class, the students come to school by train, by walking, by bus, and by car. One person rides the train. Five people walk to school. Four people ride the bus. Five people ride in cars. Most of the children walk to school because they do not live far away.

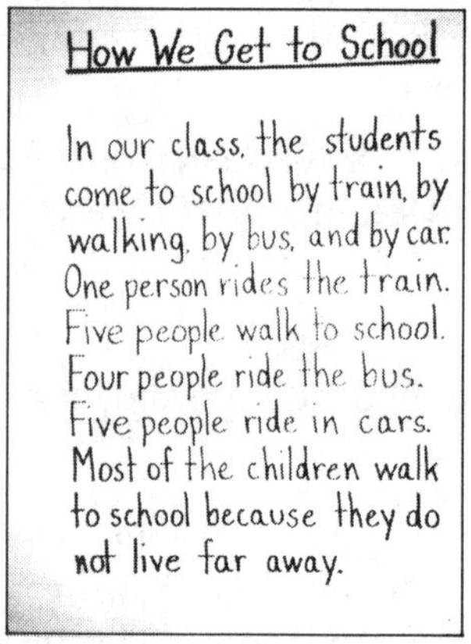

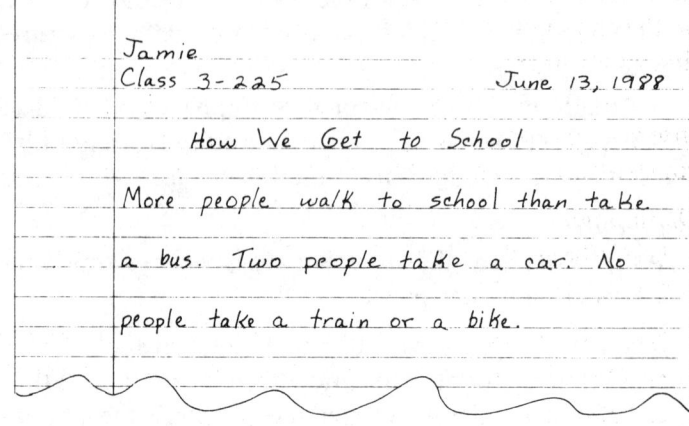

Fig. 4.3. Story by class

Jamie
Class 3-225 June 13, 1988

How We Get to School

More people walk to school than take a bus. Two people take a car. No people take a train or a bike.

Fig. 4.4. Third grader's experience story about the picture graph

ACTIVITY 5

Topic: Favorite color

Graph Form: a. Block graph
 b. Picture graph
 c. Bar graph

Graph Title: What Are Our Favorite Colors?

Objectives
1. To identify colors
2. To collect, organize, and interpret data
3. To categorize one's favorite color within a specified group
4. To construct a block graph, picture graph, and bar graph
5. To relate the results of a picture graph to a bar graph format
6. To answer comprehension questions
7. To write a story based on the graph

Vocabulary: Graph, block graph, picture graph, bar graph, review colors (if necessary), best, least

Materials
 a. Blocks (same size, one for each child); index cards cut to fit the side of the blocks; crayons (for children's favorite colors); tape to attach colored index cards to blocks; block graph labels to identify piles of blocks; space on table or desk to construct block graph for all children to see; writing tablet; marker
 b. Chalkboard; white (or light-colored) construction paper; crayons of children's favorite colors; tape; lined paper to write a story about the graph
 c. Chalkboard; chalk; lined paper to write a story about the graph

Procedure
Ask children, "Which is your favorite color?" Begin to record the children's responses in an unorganized way. "Which color do you think is the class's favorite color? How can we find out in an organized way?" Write the children's names on the board underneath a heading for the different favorite colors; make a graph.

 a. *Block graph:* Have the children color an index card using their favorite color. Tape each card to a block. Ask the children to stack their blocks in the proper pile above the appropriate label (similar to Activity 2; see fig. 2.6).

 b. *Picture graph:* Have the children color a piece of white (or light-colored) construction paper with their favorite color. Ask them to write their names on one side of the sheet and attach it in the proper row on the chalkboard. Decide which side should be showing. Perhaps the children's names should appear the first time; then place their responses on the board without the names showing.

 c. *Bar graph:* Using chalk, outline the rows (or columns) of colors posted in the picture graph activity. Insert a horizontal or vertical axis indicating the number of pieces of paper of each color. Remove the construction paper so that the outline of the bar is visible.

Questions for Discussion
1. "What is this graph about? What is a good title for this graph?" Write the title on the graph. [RBW]
2. "Which colors are listed?" [RD]
3. "Which color does the class like best?" [RBW]
4. "Which color does the class like the least?" [RBW]
5. "If you were going to package crayons for our class, which color crayon would you need more of? Why?" [RBY]

6. "Do you think boys and girls have different favorite colors or the same? How could we be sure?" [RBY]

7. "What question can you ask about this graph?"

Writing and Reading

After the children have discussed the answers to the questions, ask them to tell a story about the graphs. Using the children's words, write the story on the board or writing tablet so that all the children can see it, and encourage them to read it aloud together. By the third grade, children should be able to write their own stories and share them with others.

ACTIVITY 6

Topic: Favorite pet

Graph Form: a. Block graph
 b. Picture graph
 c. Bar graph

Graph Title: Which Pet Is Our Favorite?

Objectives
1. To identify types of animals suitable for pets
2. To collect, organize, and interpret data
3. To categorize one's favorite pet with similar pets
4. To construct a block graph, picture graph, and bar graph
5. To relate the results of a picture graph to a bar graph format
6. To answer comprehension questions
7. To write a story about the graph

Vocabulary: Pets, animals, graph, block graph, picture graph, bar graph, best, least, favorite

Materials
 a. Blocks (same size, one for each child); index cards cut to fit the side of the blocks; pencils for each child; tape; block graph labels; space on table or desk to construct block graph; writing tablet; marker
 b. Chalkboard; white (or light-colored) construction paper; pencils; tape; lined writing paper
 c. Chalkboard; chalk; lined writing paper

Procedure

Ask, "Which kinds of animals do people have as pets?" List the children's responses on the chalkboard. "If you were asked to pick one, which of these would you like best for a pet?"

 a. *Block graph:* Distribute index cards, and ask the children to draw a picture of their favorite pet and attach it to a block. Build a block graph (similar to Activity 2; see fig. 2.6).

 b. *Picture graph:* Distribute construction paper and ask the children to draw a picture of their favorite pet. Ask them to write their names on the back of their drawings and to organize the drawings on the chalkboard according to similarities. Be sure to have labels for the drawings.

 c. *Bar graph:* Using chalk, outline the rows (or columns) of pets posted in the picture graph activity. Insert a horizontal (or vertical) axis indicating the number of each type of pet. Remove the drawings so that the outline of the bar is visible.

Questions for Discussion
1. "What is this graph about? What is a good title for this graph?" Write the title on the graph so that all the children can see it. [RBW]

2. "What are the different animals that we have listed?" [RD]
3. "Which is the class's favorite pet? How do you know?" [RBW]
4. "Which pet is the class's least favorite?" [RBW]
5. "How many children can train their pets?" [RBY]
6. "How many of the pets have wings? Which ones are they? How many have four legs? Which ones are they? How many have fins? Which ones are they?" [RBY]
7. "What questions can you ask about this graph?"

Writing and Reading

After discussing the answers to the questions, ask the children to tell a story about the graph. Using the children's words, write the story on the board or writing tablet so that all the children can see it, and encourage the children to read the story aloud together. By the third grade, children should be able to write their own stories and share them with others.

ACTIVITY 7

Topic: Favorite ice cream flavors

Graph Form: a. Block graph
 b. Picture graph
 c. Bar graph

Graph Title: Which Ice Cream Flavor Is Our Favorite?

Objectives
1. To identify ice cream flavors
2. To collect, organize, and interpret data
3. To categorize one's favorite ice cream flavor with similar flavors
4. To construct a block graph, picture graph, and bar graph
5. To relate the results of a picture graph to a bar graph
6. To answer comprehension questions
7. To write a story about a graph

Vocabulary: Graph, block graph, picture graph, bar graph, ice cream flavors (review if necessary), favorite, least

Materials
 a. Blocks (same size, one for each child); index cards cut to fit the side of the blocks; crayons; tape; writing tablet; marker
 b. Chalkboard; white (or light-colored) construction paper; crayons; tape; lined writing paper
 c. Chalkboard; chalk; lined writing paper

Procedure

Ask, "What are some of the most popular flavors of ice cream?" List the names dictated by the children. You may have to limit the final list of flavors from which to choose the favorite.

a. *Block graph:* Distribute index cards for the children to color like their favorite flavors. Have them attach the index cards to the blocks and stack them according to similarities (similar to Activity 2; see fig. 2.6).

b. *Picture graph:* Distribute construction paper and ask the children to draw and color in their favorite ice cream flavors. Have them organize their drawings on the chalkboard according to similarities. Be sure to have labels for ice cream flavors.

c. *Bar graph:* Using chalk, outline the rows (or columns) of ice cream flavors posted in the picture graph activity. Insert a horizontal (or vertical) axis indicating the number of children who have each flavor as a favorite. Remove the construction paper so that the outline of the bar is visible.

Questions for Discussion
1. "What is this graph about? What is a good title for this graph?" Write the title on the graph. [RBW]
2. "Which flavors are listed?" [RD]
3. "How many children like vanilla best? Chocolate?" [RD]
4. "Which flavor is the class's favorite? Why did you say that?" [RBW]
5. "Which flavor is the class's least favorite? Why did you say that?" [RBW]
6. "If we were planning a party, which flavors should we order? Why? How should we decide?" [RBY]
7. "Do you think the other first (or second, etc.) graders like the same flavors as we do? How can we tell? How can we be sure?" [RBY] This may lead into a future activity of polling another class and comparing the results.
8. "What questions can you ask about this graph?"

Writing and Reading
After discussing the questions, ask the children to tell a story about the graph. Using the children's words, write the story on the board or writing tablet, and encourage them to read it aloud together. By the third grade, children should be able to write their own stories and share them with others.

ACTIVITY 8

Topic: Favorite color

Graph Form: Bar graph

Graph Title: Which Are Our Favorite Colors?

Objectives
1. To identify colors
2. To collect, organize, and interpret data
3. To categorize one's favorite colors according to specific groups
4. To construct a bar graph
5. To use the computer as a tool
6. To answer comprehension questions
7. To write a story about the graph

Vocabulary: Graph, bar graph, more, best, least, favorite, data

Materials: Large-box graph paper (Appendix 4); rulers, pencils, and crayons; chalkboard and chalk (or writing tablet and marker); lined paper; computers and graphing software (optional)

Procedure
Ask the children, "Which color is your favorite?" Write their responses in an unorganized fashion on the chalkboard or writing tablet. "Which color do you think is the class's favorite color? Is there a better, more organized way to present this information? [Elicit ideas from children.] How should we organize our information/data?" For example, allow children to identify their favorite colors, write their names underneath, and then convert the number of names to tally marks. Depending on the children's ability, you may want to eliminate the listing of names and tally the responses directly. See figure 8.1.

Distribute large-box graph paper. Set up axes (see fig. 8.2) and discuss labels. Insert a title (e.g., "Class 3-210's Favorite Colors"). Using the data collected from the children, identify the number of children

Our Favorite Colors

Red	*Blue*	*Green*	*Yellow*	*Purple*	*Pink*
Tyhira	Sean	Monet	Betty	Harold	Tanya
Alyson	Jose	Fran		George	
Michael	Mindy				
Juan	Terry				
	Tom				
	Ben				
	Justin				

Our Favorite Colors

Red	*Blue*	*Green*	*Yellow*	*Purple*	*Pink*
////	//// //	//	/	//	/

Fig. 8.1. Organizing data: Names and tallies

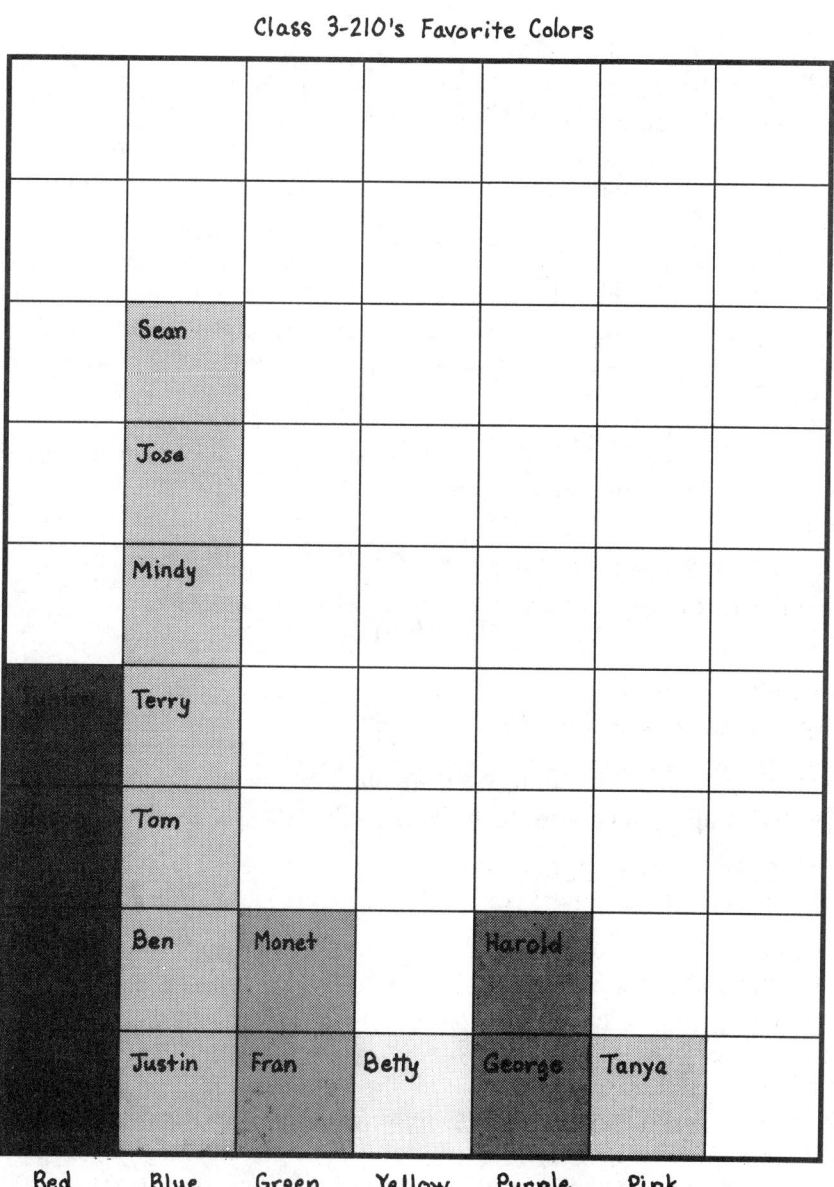

Fig. 8.2. Graphing children's favorite colors

23

selecting each color and assign one box per child in the column identified by that particular color. Have each child color the appropriate number of boxes using the particular color. Do this for each color that the children identify.

Using the Computer as a Tool

Once the data have been collected, the teacher should demonstrate the use of the computer and appropriate software for constructing a bar graph. If the children are familiar with using the computer, explain how to use the software and allow them to work in groups of two to enter the data, construct the graph, and compare the computer display with the graph they made by hand. If the software contains the capability to select options for scaling and graph formats, allow the students to explore and compare the results of these options. Depending on the availability of a printer, children should print out their work as well as save the data and the graph on a disk.

Questions for Discussion

1. "What is this graph about? How do you know this?" [RD]
2. "Which colors are listed?" [RD]
3. "How many children selected [blue] as their favorite color? [Red]? [Yellow]?" [RD]
4. "Which color does the class like best?" [RBW]
5. "Which color do children like more, [red] or [blue]?" [RBW]
6. "Which color does the class like the least?" [RBW]
7. "If you were going to package crayons for our class, which colors would you include? Why? From the information in this graph, are there any colors you might not include? Why? How can you be sure?" [RBY]
8. "Using this graph, can you tell which color the boys like best? Why? Why not? Can you tell which color the girls like best? Why? Why not? How can we find out?" Depending on the children's ability, this may lead to constructing a double bar graph with boys' and girls' results compared. [RBY]
9. "What questions can you ask about this graph?"

Writing and Reading

After the children have discussed the answers to the questions, ask them to write a story about the graph(s) and then to share their work by reading their stories aloud.

ACTIVITY 9

Topic: Favorite pet

Graph Form: a. Picture graph (using children's drawings)
b. Picture graph (using a uniform ideograph)
c. Bar graph
d. Double bar graph

Graph Title: Which Pet Is Our Favorite?

Objectives

1. To identify types of animals suitable for pets
2. To collect, organize, and interpret data
3. To categorize one's favorite pet with similar pets
4. To compare the results of a specific picture graph (using children's drawings of pets) to a more general picture graph (using a uniform ideograph) to a bar graph format
5. To construct a picture graph, bar graph, or a double bar graph

6. To use the computer as a tool
7. To answer comprehension questions based on information in a specific graph
8. To write a story based on a specific graph

Vocabulary: Graph, picture graph, bar graph, double bar graph, best, least, more, pet, animal, favorite, axes, key (or legend)

Materials

a. Drawing paper; pencils, pens, or crayons; tape; chalkboard to display drawings in picture-graph format; chalk

b. Blank Post-it notes (3" × 3", same color, one for each child to use as the uniform ideograph); chalkboard to display Post-it notes in picture-graph format; chalk; computers and graphing software

c. Large-box graph paper (see Appendix 4) or 1-cm graph paper (see Appendix 5), depending on children's ability; pencils and rulers (one for each child); computers and graphing software

d. Two different colors of Post-it notes (3" × 3", one color for girls and one color for boys); 1-cm graph paper (Appendix 5); pencils and rulers (one for each child); computers and graphing software

Procedure

Ask, "Which kinds of animals do people have as pets?" Make a list on the chalkboard of pets identified by the children. "If you were asked to pick one, which of these would you like best as a pet?"

a. *Picture graph:* Distribute drawing paper (3" × 5"), pencils, and crayons, and ask the children to draw their favorite pets (one each). Be sure to have them write their names on the back of their drawings so they can be returned. Allow the children to display their drawings (anywhere) on the chalkboard, and elicit the need for an organized way to arrange the drawings of their pets. Let the children identify the different types of pets and write these along either a horizontal or a vertical axis, labeling it "Pets." Have the children take their drawings and display them in the proper row or column. Discuss a suitable title and write it on the graph.

b. *Picture graph:* Write the different types of pets that the children identify along a horizontal or vertical axis and label it "Pets." Distribute one Post-it note to each child (all notes should be the same size and color), asking him or her to attach it in the row or column to indicate his or her favorite pet. Discuss a suitable title and write it on the graph.

c. *Bar graph:* After completing the general picture graph, draw an outline of the Post-it notes in each row (or column), and along the axis not labeled "Pets" insert numerals to represent the counting numbers indicating how many Post-it notes were used for each pet. Remove the Post-it notes, leaving the outline to represent a bar for each pet. Distribute the appropriate size of graph paper, rulers, and pencils. Have the children construct a set of axes, labeling each appropriately. Be sure to explain the importance of making each bar a uniform width. Also explain that there should be equal spacing between bars as well as equal spacing between numerals along the axis labeled "Number of Children."

d. *Double bar graph:* After asking question 7 below, children may want to examine the data differently, comparing boys' and girls' favorite pets. Begin by distributing two different colors of Post-it notes, one color for girls and one color for boys. Arrange the board so that there is enough space for two rows (or columns) of Post-it notes for each pet. Ask the children to attach their notes in the appropriate row (or column) not only according to pet but also according to the color of the note (representing boys' and girls' responses).

Distribute 1-cm graph paper, pencils, and rulers. Have the children construct a set of axes, labeling one as "Pets" and the other as "Number of Children." List the pets, leaving enough space for two bars of equal width and so that the spacing between each pet is equal. Also, be sure that the spacing between the numerals along the axis labeled "Number of Children" is equal. Have the children construct bars for boys' and girls' data. Ask them how the boys' and girls' data can be distinguished. Elicit the need for a key, or legend. Using different colors or shading on the bars for boys or girls, establish a key and insert it on the graph. Be sure to include the title of the graph.

Using the Computer as a Tool

The graphing software selected should be capable of displaying the graph forms developed in this activity. If the children are unfamiliar with how to use the software, it should be explained. Allow them to work at a computer in groups of two to enter the data, construct the graphs, and compare the computer displays with the graphs they made by hand. The software should contain options for adjusting the scale and examining alternative graph forms. The children should have time to explore the options and to discuss the results of using them. They should examine the similarities and differences among the picture graph, bar graph, and double bar graph and experiment with different scales. All graphs should be printed out so that they can be compared and discussed. The data and the graphs should be saved on a disk.

Questions for Discussion

1. "What is this graph about?" [RD or RBW]
2. "What are the different pets listed?" [RD]
3. "How many children like [dogs] the best? [Cats]? [Birds]?" [RD]
4. "Which is the class's favorite pet? How do you know?" [RBW]
5. "How many children can 'train' their pets?" [RBY]
6. "How many of the types of pets have wings? Which ones are they? How many have four legs? Which ones are they? How many have fins? Which ones are they? [RBY]
7. "From this graph, can you tell which pet boys like best? Can you tell which pet girls like least?" Before constructing a double bar graph, ask, "How can we find out?" [RBY] After constructing a double bar graph, ask, "How do you know this?" [RBW]
8. "What questions can you ask about this graph?"

Writing and Reading

After the children have discussed the answers to the questions, ask them to write a story about the graph(s). Then ask them to share their work by reading their stories aloud. Encourage the children to question and criticize their peers constructively.

ACTIVITY 10

Topic: Number of children in your family

Graph Form: a. Picture graph (using a uniform ideograph and one-to-one correspondence)
b. Picture graph (using a uniform ideograph and two-to-one correspondence with key)

Graph Title: How Many Children Are in Your Family?

Objectives

1. To identify the number of children in one's family
2. To collect, organize, and interpret data
3. To categorize the number of children in one's family with other children's responses
4. To construct a picture graph using symbols when the number of children and symbols are in one-to-one correspondence and two-to-one correspondence
5. To compare the results of a picture graph in which the number of children and the ideographs are in one-to-one correspondence with one in which the number of children and the ideographs are in two-to-one correspondence
6. To use the computer as a tool
7. To answer comprehension questions based on information in a specific graph
8. To write a story based on a specific graph

Vocabulary: Graph, picture graph, axes, key (or legend)

Materials: Uniform "smiley" faces prepared on Post-it notes (3" × 3", all the same color); one pair of scissors; pencils (one for each child); worksheet for each child (see Appendix 16); chalkboard and chalk

Procedure

Ask the children, "How many brothers and sisters do you have? How many children are in your family?" List some answers on the chalkboard. Try to identify the least number of children and the most number of children. Prepare a horizontal axis on the chalkboard, labeled "Number of Children," and say, "We are going to represent the number of children in our families by using a picture graph."

Distribute a "smiley" face to each student. Ask them to place the Post-it note above the numeral that represents the number of children in their family. Using the left side of the worksheet in Appendix 16, ask the students to copy the picture graph from the chalkboard. Be sure to label the horizontal axis and identify a title for the graph. Discuss questions 1–5.

Now mention that we would like to decrease the number of Post-it notes that we need for this graph: "How can we use fewer notes?" Elicit that we can let each symbol represent a multiple number of students in our class. For example, let each symbol represent two students. "If two symbols represent two students in Graph 1, how many symbols will be needed to represent two students in Graph 2?" Beginning with the first entry of the horizontal axis on the chalkboard, ask the class to convert each column so that one symbol represents two students. For an odd number of students, use a pair of scissors to cut a smiley face in half. Be sure to have the students insert a key (or legend) for each graph. Using the right side of the worksheet in Appendix 16, ask the students to copy the modified picture graph from the chalkboard. Be sure to label the horizontal axis and identify a title for the graph. Ask the students to explain the graph (see fig. 10.1).

Fig. 10.1

Using the Computer as a Tool

Select graphing software that has a picture graph capability with the option of creating a legend to adjust the number of items each ideograph can represent. Once the children are familiar with how to use the software, allow them to work at computers in groups of two to enter the data, construct the graphs, and compare the computer displays with the graphs they made by hand. The pupils should discuss the similarities and differences between picture graphs that represent one-to-one correspondence and many-to-one correspondence for the same data. If a printer is available, the graphs should be printed out. The data and the graphs should be stored on a disk.

Questions for Discussion

1. "What is this graph about?" [RD or RBW]

2. "How many students in our class have two children in their families? Three children? Four children?" [RD]

3. "How many children are in the most number of families?" [RBW]
4. "How many children are in the fewest number of families?" [RBW]
5. "How many students in our class are the oldest child in their families?" [RBY]

Encourage students to compare Graph 1 and Graph 2:

6. "How are the two picture graphs the same? How are they different?" [RBY]
7. "What questions can you ask about these graphs?"

Writing and Reading

After the students have discussed the answers to the questions, ask them to write a story about the graph(s) (see fig. 10.2). Ask them to share their work by reading their stories aloud, and encourage them to question and criticize each other constructively.

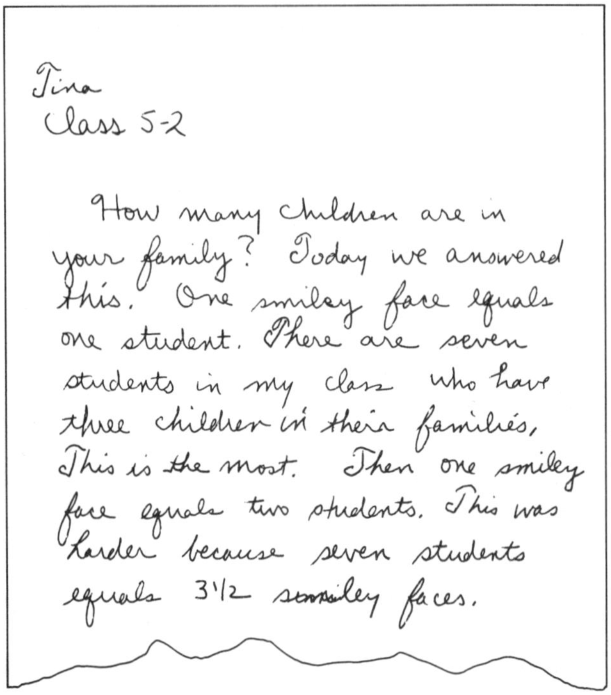

> Tina
> Class 5-2
>
> How many children are in your family? Today we answered this. One smiley face equals one student. There are seven students in my class who have three children in their families. This is the most. Then one smiley face equals two students. This was harder because seven students equals 3 1/2 smiley faces.

Fig. 10.2. A fifth grader's experience story about the graph activity

ACTIVITY 11

Topic: Favorite ice cream flavor

Graph Form: a. Picture graph (using uniform ideographs and one-to-one correspondence)
  b. Picture graph (using uniform ideographs and two-to-one correspondence with key)
  c. Bar graph
  d. Double bar graph

Graph Title: Which Ice Cream Flavor Is Your Favorite?

Objectives

1. To identify ice cream flavors
2. To collect, organize, and interpret data
3. To categorize one's favorite ice cream flavor with similar flavors
4. To construct a picture graph, bar graph, and double bar graph
5. To compare the results of a picture graph in which the number of children and the ideographs are in one-to-one correspondence with one having two-to-one correspondence
6. To use the computer as a tool
7. To compare the results of the class survey, displayed in a bar graph and employing the use of percents, with the results of a national survey published in a newspaper

8. To answer comprehension questions based on information in a specific graph
9. To express the meaning of a graph in prose

Vocabulary: Graph, picture graph, bar graph, double bar graph, axes, key (or legend), percent, representative

Materials

a. Uniform symbols of ice-cream cones, made from a stencil or on a photocopying machine; chalkboard to display symbols; tape; large-box graph paper (Appendix 4) for each child; pencils and rulers for each child; computers and graphing software

b. Same as (a); scissors

c. 1-cm graph paper (Appendix 5), pencils, and rulers for each child; computers and graphing software

d. Same as (c)

Procedure

Ask, "What are some of the most popular flavors of ice cream?" List the names dictated by the children. You may have to limit the final list of flavors from which the children may choose their favorites. To compare class results to a national survey, give the children only the choices from the survey. See figure 11.1. Prepare a set of axes on the chalkboard, and label the vertical axis "Ice Cream Flavors."

a. *Picture graph (one-to-one correspondence):* Distribute a picture of an ice-cream cone to each student. Have each pupil tape the symbol in the row that represents his or her favorite flavor. Distribute large-box graph paper so that the students can copy the graph from the chalkboard, drawing in the uniform ice-cream cone symbols. Discuss a possible title and insert it on the graph. Discuss questions 1–6 below.

b. *Picture graph (two-to-one correspondence):* Now mention that we would like to decrease the number of symbols used in this graph: "How can fewer symbols be used?" Elicit that we can let each symbol represent a multiple number of students in the class. For example, let each symbol represent two students. "If one symbol represents one student in our first graph, how many symbols will be needed to represent two students in this graph?" (Remember, one symbol represents two pupils.) "How many symbols will be needed to represent five pupils?"

Beginning with the top entry of the vertical axis on the chalkboard, ask the class to convert each row so that one symbol represents two students. For an odd number of students, use a pair of scissors to cut a cone in half. Distribute more large-box graph paper and have the students convert the one-to-one picture graph to a two-to-one picture graph. Be sure that they insert a key (or legend) for each graph. Ask them to explain each graph. Discuss question 7 below.

c. *Bar graph:* Distribute 1-cm graph paper, rulers, and pencils. On the chalkboard list the following flavors: vanilla, chocolate, neapolitan, vanilla fudge, and cookies and cream (from the national survey reported in fig. 11.1). Ask the students to vote for the flavor they like best. Use tally marks on the chalkboard to indicate the number of students who prefer each flavor. Ask the students to construct a bar graph from these data. Be sure that they label each axis and insert an appropriate title for the graph.

Fig. 11.1 (Copyright 1987, *USA Today*. Reprinted with permission.)

Ask the students to express the class results in terms of percent so that they can be compared with the results of the national survey (see fig. 11.1). Discuss question 8 below.

d. *Double bar graph:* Depending on whether you decide to have the survey open-ended (i.e., allow students to state their favorite flavors) or closed (i.e., allow a limited number of choices from which they can select), ask the students to collect and analyze the data for boys and girls separately. A poll can be conducted by students and tally marks can be displayed on the chalkboard. Distribute 1-cm graph paper, rulers, and pencils, and ask the students to construct a double bar graph. Remind them to make bars with equal widths and to leave equal spacing between flavors and numerals. Also, make sure they identify a title and insert it on the graph.

Using the Computer as a Tool

The graphing software selected should be capable of displaying the graph forms developed in this activity. Once the children are familiar with the software, allow them to work at the computers in groups of two to enter the data, construct the graphs, and compare the computer displays with the graphs they made by hand. The software should contain options for adjusting the scale and examining alternative graph forms. The children should have time to explore these options and to discuss the results of using them. They should examine the similarities and differences among the picture graph, bar graph, and double bar graph and experiment with different scales.

To facilitate a comparison of the class's data with those from the national survey, use the computer to convert the bar graph to a circle graph, if the software provides percent equivalents.

All graphs should be printed out so that they can be compared and discussed. The data and the graphs should be stored on a disk.

Questions for Discussion

1. "What is this graph about?" [RD or RBW]

2. "How many students like vanilla best? Chocolate? Chocolate-chip mint?" [RD]

3. "Which flavor does the class like the best? How do you know this?" [RBW]

4. "Which flavor is the least popular?" [RBW]

5. "If we were going to plan a class party and we wanted to serve ice cream, how many flavors would you decide to order? Why? Which flavors would they be?" [RBY]

6. "What questions can you ask about this (these) graph(s)?"

For activities (a) and (b) ask the following:

7. "How are the two picture graphs the same? How are they different?" [RBY]

For activity (c) ask:

8. "How do our class results compare with the results of the national survey? What questions should we ask about the results reported in the national survey? Do you think it's representative of the entire U.S. population? Why? Why not? Do you think it's fair to compare the two graphs? Why? Why not?" [RBY]

In preparation for activity (d), ask:

9. "Is vanilla more popular among the boys or among the girls in our class? How do you know this? Could you tell this from the bar graph [activity (c)]? [RBW] How could we find out?

After activity (d), ask:

10. "On the basis of the results in this graph, would you change your answer to question 5? If not, why not? If yes, how? Why?" [RBW]

Writing and Reading

After the students have discussed the answers to the questions, ask them to write a story about the graph(s). Ask them to share their work by reading their stories aloud. Encourage them to question their peers.

ACTIVITY 12

Topic: Favorite game (board game, card game, athletic or sports game, etc.)

Graph Form: a. Bar graph
b. Double bar graph

Graph Title: Which Is Your Favorite Game?

Objectives
1. To identify different types of games
2. To collect, organize, and interpret data
3. To categorize one's favorite game with similar responses
4. To construct a bar graph and a double bar graph
5. To use the computer as a tool
6. To answer comprehension questions based on information in a specific graph
7. To express the meaning of a graph in prose

Vocabulary: Graph, bar graph, double bar graph, legend (or key), horizontal, vertical, survey, poll, data, compare, analyze

Materials
a. Chalkboard and chalk (or overhead projector, transparencies, and markers); 1-cm graph paper (Appendix 5), pencils and rulers for each pupil; survey sheet (similar to Appendix 10); computers and graphing software
b. Same as (a)

Procedure
Using a survey form designed by the students, ask them to take a survey of favorite games. The survey can be conducted either in the classroom or as an outside assignment. The pollsters must be sure that they do not collect data from the same student more than once. To construct a double bar graph, they can collect data separately from boys and girls. To construct a bar graph, the data can be combined; to construct a double bar graph, the data can be graphed separately with different colors or shadings.

a. *Bar graph:* Once the data have been collected, discuss with the students how to organize a bar graph (or if they prefer, a picture graph is also suitable). Distribute 1-cm graph paper and have the students draw a set of axes. This should also be done on the chalkboard (or on a transparency). Have the students determine whether the bar graph will be horizontal or vertical. Depending on the number of pieces of data, it might be necessary to use multiples of 2 or 5 (or some other multiple) along the axis labeled "Number of Students." Discuss how to write the numerals along the axis (i.e., stress the meaning of each box and each line). Discuss questions 1–5 below.

b. *Double bar graph:* If the data are collected for boys and girls separately, the students can construct a double bar graph. Have the results displayed on the chalkboard or overhead transparency. Distribute 1-cm graph paper, and ask the students to decide whether the bars will be horizontal or vertical. Have them label the axes accordingly. Be sure that the bars have equal widths and that the spacing between "Names of Games" on one of the axes is equal, as well as the spacing between the numerals listed for the "Number of Students." Have the students color or shade the bars and include a legend. They should also identify an appropriate title and insert it on the graph. Discuss questions 6–8 below.

Using the Computer as a Tool
Select graphing software that can be used to construct a bar graph and a multiple bar graph. Once the children are familiar with the software, allow them to work at computers in pairs to enter the data, construct the graphs, and compare the computer displays with the graphs made by hand. The software

should contain options for adjusting the scale and examining alternative graph forms. Give children time to explore the options and to discuss the results of using them. If a printer is available, have them print out the graphs to facilitate comparisons. The data and the graphs should be saved on a disk.

Questions for Discussion
1. "What is this graph about?" [RD or RBW]
2. "What are the different games listed?" [RD]
3. "How many students participated in the survey?" [RBW]
4. "Which game is the most popular?" [RBW]
5. "If you were planning to have a party, how many games would you consider playing at your party? Why? Which games would you select?" [RBY]
6. "Which game is the most popular among the boys? Among the girls?" [RBW]
7. "If you analyze boys' and girls' responses separately, would you answer question 5 differently? Why? Why not?" [RBY]
8. "What questions can you ask about these graphs?"

Writing and Reading
After the students have discussed the answers to the questions, ask them to write a description of the graph (i.e., what it means to them). Ask them to share their work by reading their stories aloud, and encourage them to question each other.

ACTIVITY 13

Topic: Favorite leisure activity

Graph Form: Double bar graph

Graph Title: What Is Your Favorite Leisure Activity?

Objectives
1. To identify activities pursued during one's leisure time
2. To collect, organize, and interpret data
3. To categorize one's favorite leisure activity with similar responses
4. To construct a double bar graph
5. To compare the results of a class (or school survey) with the results of a national survey published in a newspaper
6. To answer comprehension questions based on information in a specific graph
7. To express the meaning of a graph in prose

Vocabulary: Graph, double bar graph, horizontal, vertical, survey, poll, compare, analyze, percent, representative

Materials: Chalkboard and chalk (or overhead projector, transparencies, and markers); 1-cm graph paper (Appendix 5) or 1/4" graph paper (Appendix 6), pencils, and rulers for each student; survey sheet (similar to Appendix 10); graph from a national survey (see fig. 13.1)

Procedure
Ask the students, "What are some activities that people pursue during their leisure time?" List their ideas on the chalkboard or overhead transparency, and compare their list with the activities listed in a national survey (see fig. 13.1). The student survey should be based on the choices offered in the national survey so that their results can be compared.

Fig. 13.1 (Copyright 1987, *USA Today*. Reprinted with permission.)

The survey can be conducted either in the classroom or as an outside assignment. The pollsters must be sure that they do not collect data from the same student more than once. To construct a double bar graph, data collected from boys and girls should be kept separate.

Once the data have been collected, discuss with the students how to organize a double bar graph. Distribute graph paper and have the students draw a set of axes. This should also be done on the chalkboard (or a transparency). Have the students decide whether the double bar graph will be horizontal or vertical. Depending on the number of pieces of data, it might be necessary to use multiples of 2 or 5 (or some other multiple) along the axis labeled "Number of Students." Discuss how to write the numerals along the axis, stressing the meaning of each box and each line.

Once the graph is completed, convert the results to percents so that the graph constructed by the students can be compared with the graph in figure 13.1.

Questions for Discussion

1. "What is this graph about?" [RD or RBW]
2. "What are the leisure activities listed?" [RD]
3. "Are there any other leisure activities that you think should have been included in the survey? If so, what are they?" [RBY]
4. "Why did we restrict the number of leisure activities included in our survey?" [RBY]
5. "What is the most popular leisure activity for boys? For girls? How do these results compare with those of men and women in the newspaper graph [fig. 13.1]?" [RBW]
6. "What questions should we ask about the results reported in the national survey? Do you think it's representative of the entire U.S. population? Why? Why not?" [RBY]
7. "Do you think it's fair to compare the class's results with the national survey results? Why? Why not?" [RBY]

Writing and Reading

After the students have discussed the answers to the questions, ask them to write a description of the graph (i.e., what it means to them). Ask them to share their work by reading their stories aloud, and encourage them to question each other.

ACTIVITY 14

Topic: Rooms with a television

Graph Form: Bar graph

Graph Title: Rooms in Our Houses and Apartments Containing a TV

Objectives
1. To identify rooms in houses and apartments that have a television
2. To collect, organize, and interpret data
3. To categorize rooms according to similar responses
4. To construct a bar graph
5. To compare the results of a class (or school or neighborhood) survey with the results of a national survey published in a newspaper
6. To answer comprehension questions based on information in a specific graph
7. To express the meaning of a graph in prose

Vocabulary: Graph, bar graph, horizontal, vertical, survey, poll, compare, analyze, percent, representative

Materials: Chalkboard and chalk (or overhead projector, transparencies, and markers); 1-cm graph paper (Appendix 5) or 1/4" graph paper (Appendix 6), pencils, and rulers for each student; survey sheet (similar to Appendix 10); graph from a national survey (see fig. 14.1)

Procedure

Ask students, "On the average, how many televisions do you think people have in their houses or apartments? In which rooms do you think people have televisions?" To compare the data collected by students with the data collected in the national survey, use the following rooms: living room, master bedroom, family room, kitchen, children's bedroom, and den. See figure 14.1. "How could we go about collecting the answers to these questions?" Conduct a survey. "What would be an efficient way of organizing and reporting the results of our survey?" Depending on students' ability, you can have them design their own survey form (see Appendix 10 for some ideas). The survey can be conducted either in the classroom or as an outside assignment. As the students design a plan to conduct a survey, have them discuss other related questions they might want to ask. Remind them to be sure that they do not collect data from the same respondent more than once.

Fig. 14.1 (Copyright 1987, *USA Today*. Reprinted with permission.)

Once the data have been collected, discuss with the students how to organize the bar graph. Distribute graph paper and have the students daw a set of axes. This should also be done on the chalkboard (or on a transparency). Have the students decide whether the bar graph should be horizontal or vertical. Depending on the number of pieces of data, it might be necessary to use multiples of 2 or 5 (or some other multiple) along the axis labeled "Number of Students" (or Respondents). Review how to write the numerals along the axis—that is, stress the meaning of each box and each line. Be sure that the students identify a title and insert it on their graphs.

Once the graph is completed, convert the results to percents so that the graph constructed by the students can be compared with the graph of the national survey (fig. 14.1).

Questions for Discussion
  1. "What is this graph about?" [RD or RBW]
  2. "How many people have a television in their living room? Master bedroom? Children's bedroom?" [RD]
  3. "Are there more homes with televisions in the living room or in the kitchen? Why do you think this is so?" [RBW and RBY]
  4. "Which room seems to be the most popular room for viewing television? Why?" [RBW and RBY]
  5. "On the average, how many televisions are there in each house? Since multiple answers were given by individual respondents, can this really be answered using only the data in the bar graph? How can the data be collected and organized so that we could answer this question?" [RBY]
  6. "What questions should we ask about the results reported in the national survey? Do you think it's representative of the entire U.S. population? Why? Why not?" [RBY]
  7. "Do you think it's fair to compare the class's results with the national survey results [fig. 14.1]? Why? Why not?" [RBY]
  8. "Why do you think televisions are found in so many different rooms in our houses and apartments?" [RBY]
  9. "Do you think there are any problems that may arise from having televisions in so many different rooms? If so, what are they?" [RBY]
  10. "What questions can you ask about these graphs?"

Writing and Reading
After the students have discussed the answers to the questions, ask them to write a description of the graph (i.e., what it means to them). Ask them to share their work by reading their stories aloud, and encourage them to question each other.

ACTIVITY 15

Topic: Height

Graph Form: Bar graph

Graph Title: How Tall Are You?

Objectives
  1. To determine one's height by measuring in metric units of measure
  2. To collect, organize, and interpret data
  3. To compare the heights of students working in a small group (i.e., 4–5 students)
  4. To construct a bar graph
  5. To use the computer as a tool
  6. To answer comprehension questions based on information in a specific graph
  7. To express the meaning of a graph in prose

Vocabulary: Graph, bar graph, height, centimeters, meters, metric tape measure, compare

Materials: Approximately ten metric tape measures (two for each of five groups) attached to wall; recording sheets (similar to Appendix 11); 1-cm graph paper (Appendix 5) or 1/4" graph paper (Appendix 6), pencils, and rulers for each student; computers and graphing software

Procedure

Ask, "Who is the tallest in our class? How can we be sure?" (By standing next to each other; by measuring heights.) Allow students to work in groups of four or five to measure each other's height and construct a four- or five-item bar graph. Attach metric tape measures to the wall in several places around the classroom. You will need to attach two tape measures end to end to measure students' heights. Review how to read a tape measure and how to add on the amount from the second tape measure. Show the students doing the measuring how to use rulers as a guide, placing them gently on the head of the student being measured. Each student should record the heights of all the members in his or her group.

Once the data have been collected, the students in each group should discuss how to organize the bar graph—the bars should be organized vertically so that they visually represent height. Distribute graph paper and have the students draw and label a set of axes. The students should decide which multiples to use so that everyone's height can fit on the graph properly. Remind the students that equal spacing is necessary between bars and between numerals listed along the vertical axis. Students should identify a title and insert it on the graph.

Using the Computer as a Tool

Once their heights have been recorded, students should work in pairs at the computer to enter their group's data, construct a bar graph, and explore how the graph changes as different scales and alternative graph forms are used. Students should report the results of their explorations to the rest of the class. Graphs that seemingly distort the data and those that represent the data fairly should be discussed. They should print out their graphs and save the data and the graphs on a disk.

Questions for Discussion

1. "What is this graph about?" [RD or RBW]
2. "How tall is [insert name]?" [RD]
3. "Who is the tallest in your group?" [RBW]
4. "Who is the shortest in your group?" (For some children this may be a sensitive question.) [RBW]
5. "How much taller is [insert name] than [insert name]?" [RBW]
6. "From the information in this graph, can you guess who weighs the most?" (This, too, may be a sensitive question.) "Why did you answer that way?" [RBY]
7. "From the information in the graph, who do you think wears the largest shoe? The smallest?" [RBY]
8. "Think of a question you can ask about this graph."

Writing and Reading

After the students have discussed the answers to the questions, ask them to write a description of the graph (i.e., what it means to them). Ask them to share their work by reading their stories aloud or switching papers among groups. Allow the group members to compare the graphs with the written descriptions of them. Allow the students to ask each other questions about their written statements.

ACTIVITY 16

Topic: Daily schedule of activities

Graph Form: Circle graph

Graph Title: How Do You Spend a Typical Saturday?

Objectives

1. To identify the amount of time spent on different activities during a typical Saturday
2. To collect, organize, and interpret data
3. To construct a circle graph on a stenciled circle partitioned into twenty-four congruent sections

4. To construct a circle graph using the computer as a tool
5. To answer comprehension questions based on the information in a specific graph
6. To express the meaning of a graph in prose

Vocabulary: Graph, circle graph, fractional part

Materials: Data sheet and circle graph outline (Appendix 8) for each student; pencils and crayons; computers and graphing software

Procedure

Ask, "How do you spend a typical Saturday? What are some activities you enjoy doing? How long does each activity take?" To examine activities by the hour, have the students list their activities and the time spent doing them on a data sheet. Once the data have been entered on the sheet, distribute the outline of the circle graph. Students should use each section as an hour and insert their activities appropriately. (See fig. 16.1.)

Using the Computer as a Tool

Select graphing software that has the capability of constructing circle graphs and expressing results in terms of equivalent percents. Once the students are familiar with how to use the software, allow them to work in pairs to enter their data, construct a circle graph, and explore different software options. The students should print out the graphs they design to facilitate making comparisons. The data and the graphs should be stored on a disk.

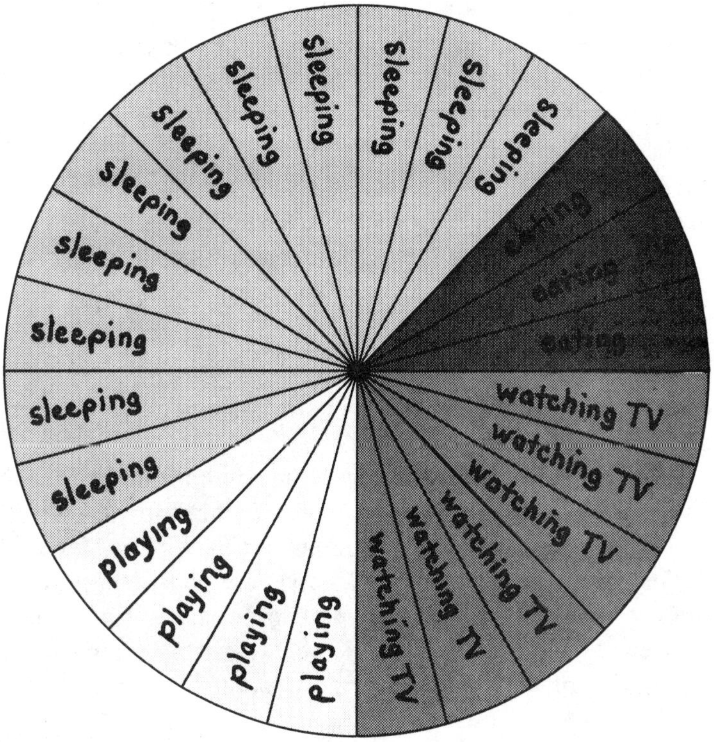

Questions for Discussion

1. "What is this graph about?" [RD or RBW]

2. "How much time do you spend watching television? Eating? Sleeping?" [RD]

3. "For which activity do you spend the least amount of time? The most amount of time?" [RBW]

Fig. 16.1. Sample of a sixth grader's circle graph

4. "How might this graph be different for a weekday during the school year? A summer day?" [RBY]

5. "For which activities do you spend a total of more than three hours?" [RD]

6. "For which activities do you spend a total of less than three hours?" [RD]

7. "What fractional part of a Saturday do you spend sleeping? Watching TV? Reading?" [RBY]

8. "After constructing a graph to analyze how you spend a typical Saturday, are there any activities that you think you spend too much time on? Not enough time on? What are they? How would you change these?" [RBY]

9. "What question can you ask about this graph?"

Writing and Reading

After the students have discussed the answers to the questions, ask them to write a description of the graph (i.e., what it means to them). Ask them to share their work by switching papers, data sheets, and graphs. Allow the students to analyze each other's work, encouraging them to question and criticize constructively.

ACTIVITY 17

Topic: Spending daily allowance

Graph Form: Circle graph

Graph Title: How Do You Spend Your Daily School Allowance?

Objectives
1. To identify the amount of money available for spending on a school day
2. To collect, organize, and interpret data
3. To apply knowledge of ratio and proportion for constructing a circle graph
4. To use a compass and protractor to construct a circle graph
5. To use the computer as a tool
6. To answer comprehension questions based on the information in a specific graph
7. To express the meaning of a graph in prose

Vocabulary: Graph, circle graph, degrees, compass, protractor, ratio, proportion, percent, fractional part, allowance, typical, central angle

Materials: Overhead projector, transparencies, and markers (or chalkboard and chalk); protractor and compass for the overhead projector (or for the chalkboard); compasses, protractors, unlined paper for each student; computers and graphing software

Procedure

Ask, "What are some of the items you buy on a typical school day? How much money do you need for school? How much money do your parents or guardians give you for a typical school day? What is an allowance?"

Have the students write down the total amount of money they receive for a daily allowance and how they spend it, listing the items they buy and how much they cost. Have them determine what fractional part of the total allowance is spent for each item (for an example, see fig. 17.1).

Distribute unlined paper, compasses, and protractors to each student. (If necessary, review safety procedures for using compasses.) Review the characteristics of a circle (e.g., it contains 360 degrees; it is a set of points equidistant from one point called the center). Have the students set a radius on their compasses (e.g., 3 inches) and draw a circle. Ask such questions as, "If we want to draw one-fourth of the circle, how many degrees would we mark off? [90 degrees] If we want to draw one-half of the circle, how many degrees would we mark off? [180 degrees] If we want to draw one-third of the circle, how many degrees would we mark off? [120 degrees] How can we calculate the number of degrees to mark off when we know the fractional part of the circle that we want to draw?" [Multiply 360 degrees by the fractional part or make a proportion; see below.]

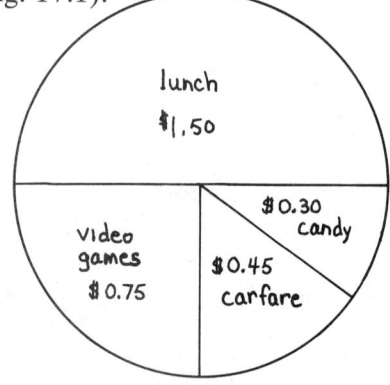

Fig. 17.1. Sample of a seventh-grader's circle graph

"Look at the first item on your list that you buy using your daily school allowance. What part of your allowance do you use to buy it?" Have the students multiply 360 degrees by this fractional amount to compute the number of degrees to mark off in the circle. Review how to construct central angles in the circle with a protractor. Have the students complete their circle graphs by computing fractional parts of the circle and constructing the proper central angles. Have them identify the number of degrees for the fractional parts on the side of the graph. Students should indicate the item and the amount of money for each item (or the fractional part or percent of the total allowance) in each section of the circle graph. Be sure to have the students identify a title for the graph.

Some students may be able to compute the number of degrees in the central angles to be constructed by setting up a proportion. For example, for lunch (in fig. 17.1),

$$1.50/3.00 = x/360.$$

Using the Computer as a Tool

The graphing software selected should provide options for constructing a circle graph and expressing the results in terms of equivalent fractional parts or percents. Once the students are familiar with how to use the software, allow them to work in pairs to enter their data, construct a circle graph, and explore the different software options. They should try to represent the data in different graph forms and describe whether the different forms are appropriate, explaining the criteria they used. To facilitate making comparisons, students should print out their graphs. The data and the graphs should be saved on a disk.

Questions for Discussion

1. "What is this graph about?" [RD or RBW]
2. "How much money do you spend for lunch? For carfare?" [RD]
3. "For which item do you spend [insert amount of money]?" [RD]
4. "What part of your allowance do you spend on lunch? Carfare?" [RBW]
5. "How can you budget your money so that you can save some [or more] of it? What nonessential items can you eliminate from your shopping list to save money? Would you want to eliminate them?" [RBY]
6. "If you were given an extra dollar a day, what would you do with it?" [RBY]
7. "During a five-day school week, about what part of your five-day allowance is spent on lunch?" [RBY]
8. "Think of a question you can ask about this graph."

Writing and Reading

After the students have discussed the answers to the questions, ask them to write a description of the graph (i.e., what it means to them), and have them share their work by switching papers and graphs. Allow them to analyze each other's work, encouraging them to question and criticize constructively.

ACTIVITY 18

Topic: Types of dwellings

Graph Form: a. Bar graph
b. Circle graph

Graph Title: In What Type of Dwelling Do You Live?

Objectives

1. To identify the types of dwellings in which people may live
2. To collect, organize, and interpret data
3. To categorize one's dwelling place according to similar responses
4. To construct a bar graph and a circle graph
5. To use the computer as a tool
6. To convert the information presented in a bar graph into a circle graph format
7. To compare the type of information that can be conveyed in a bar graph with the type that can be conveyed in a circle graph
8. To compare the results of a class (or school or neighborhood) survey with the results of a national survey published in a newspaper

9. To answer comprehension questions based on the information in a specific graph
10. To express the meaning of a graph in prose

Vocabulary: Graph, bar graph, circle graph, dwelling, percent, representative

Materials: Chalkboard and chalk (or overhead projector, transparencies, and markers); 1/4" graph paper (Appendix 6) or 5-mm graph paper (Appendix 7), unlined paper (for circle graph), pencils, rulers, compasses, and protractors for each child; a graph from a national survey; computers and graphing software

Procedure

Ask, "In what type of dwelling do you live?" List students' responses on the chalkboard. To compare the data collected by students with the data collected in the national survey, use the following types of dwellings: single-family house, apartment, two-family house, co-op or condominium, trailer home, town house. See figure 18.1. "How could we go about collecting information about the types of dwellings in which people live? What would be an efficient way of organizing and reporting the results of our survey?"

The survey can be conducted either in class or as an outside assignment. Remind the students to be sure that they do not collect data from the same respondents more than once.

a. *Bar graph:* Once the data have been collected, discuss with the students how to organize the bar graph. Distribute graph paper and have them draw a set of axes. This should also be done on the chalkboard or on a transparency. Have the students decide whether the bar graph should be horizontal or vertical. Depending on the number of pieces of data, it might be necessary to use multiples of 2 or 5 (or some other multiple) along the axis labeled "Number of Students" (or Respondents). Review how to write numerals along the axis, stressing the meaning of each box and each line.

Once the graph is completed, have students convert the results to percents so that the graph constructed by the students can be compared with the graph in figure 18.1. See figure 18.2.

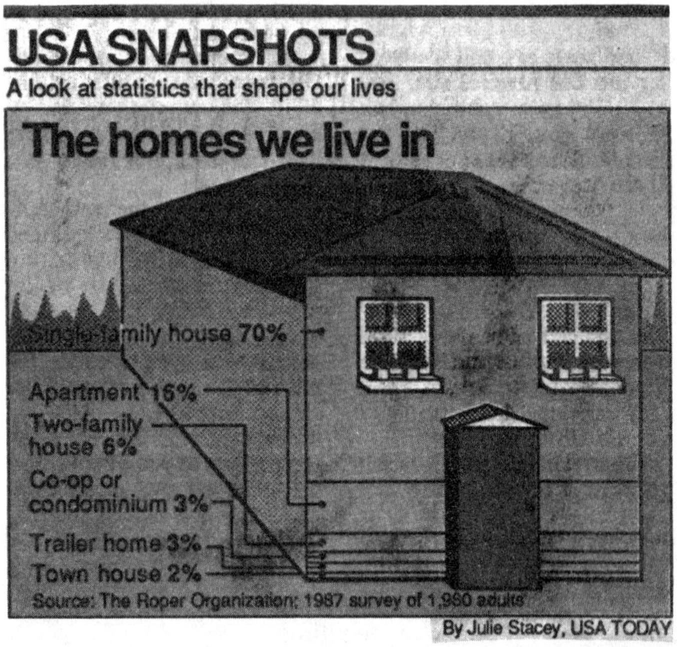

Fig. 18.1 (Copyright 1987, *USA Today*. Reprinted with permission.)

b. *Circle graph:* Distribute unlined paper, protractors, and compasses to each student. (Remind them about the safety procedures for using the compass.) Have students determine the central angle measures for converting the bar graph to a circle graph (see Activity 17). Using the protractor, each student should draw sectors of the circle to represent the percentage of dwellings of each type (see fig. 18.3). Be sure they label each section of the circle graph and identify a title for the graph.

Using the Computer as a Tool

Select graphing software that can be used to convert data expressed in a bar graph to a circle graph. It would be helpful if there is an option to express the results of the circle graph in equivalent percents.

Allow the students to work at the computers in pairs to enter their data, to construct the bar and circle graphs, and to explore the other software options available. Students should print out their graphs so they can make comparisons. Discuss graphs that seem to distort the data and those that represent the data fairly. The data and the graphs should be saved on a disk.

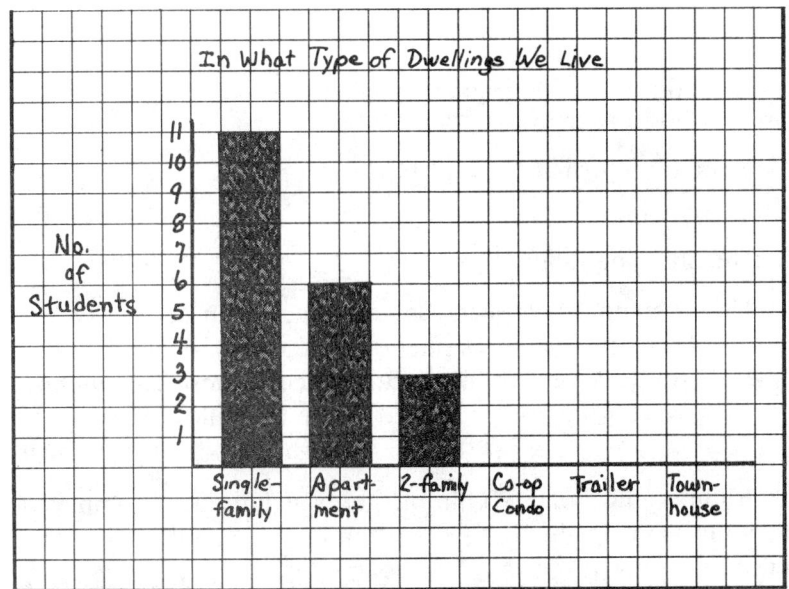

Fig. 18.2. Sample of an eighth-grader's bar graph

Fig. 18.3. Sample of an eighth-grader's circle graph

Questions for Discussion
1. "What is this graph about?" [RD or RBW]
2. "How many students live in single-family houses? Apartments? Trailer homes?" [RD]
3. "In the graph from the class survey, are there more students who live in two-family houses or in apartments?" [RBW]
4. "In the graph from the national survey, are there more people who live in two-family houses or in apartments?" [RBW]
5. "What questions should we ask about the results reported in the national survey? Do you think it's representative of the entire U.S. population? Why? Why not?" [RBY]
6. "Do you think it's fair to compare the class's results with the national survey results [fig. 18.1]? Why? Why not?" [RBY]
7. "What is the difference between the information displayed in a bar graph and the information displayed in a circle graph? What is your preference and why?" [RBY]
8. "What question can you ask about these graphs?"

Writing and Reading
After students have discussed the answers to the questions, ask them to write a description of the graphs (i.e., what they mean to them). Ask them to share their work by reading their stories aloud, and allow them to analyze each other's work. Encourage them to question and criticize constructively.

ACTIVITY 19

Topic: A comparison of height and standing long jump distance

Graph Form: Double bar graph

Graph Title: Can You Jump Your Height?

Objectives
1. To determine one's height by measuring in English units of measure
2. To determine the distance one is able to jump

3. To collect, organize, and interpret data
4. To construct a double bar graph
5. To compare the height and distance jumped for a small group of students
6. To use the computer as a tool
7. To answer comprehension questions based on information in a specific graph
8. To express the meaning of a graph in prose

Vocabulary: Graph, double bar graph, height, standing long jump, compare

Materials: Approximately ten English-unit (i.e., standard) tape measures attached in sets of two; a standing long jump mat (see fig. 19.1); 1/4" graph paper (Appendix 6) or 5-mm graph paper (Appendix 7), pencils, rulers, and data sheets (Appendix 11) for each student; computers and graphing software

Procedure

Ask the students, "Do you think you can jump your height? One of the expectations of a common physical fitness test is that you be able to jump your height. Let's see whether you can do this." This activity should be fun; the results should not be used for a physical fitness grade.

Allow the students to work in small groups of four or five so that they can measure each other's height (see Activity 15) and distance jumped. Review how to read a tape measure and how to add on the amount from the second tape measure. Show the students how to measure each other's height by placing a guide (such as a ruler) gently on top of the head of the student being measured (see fig. 19.2). Each student should record the heights of all the members of his or her group, using a data sheet similar to Appendix 11.

A standing long jump mat is usually marked in feet. If a mat is not available, mark off a distance of seven feet on the floor in one-foot intervals. (Be advised that not using a mat may be dangerous, depending on the floor surface.) Each student should record the distance jumped for all the members of his or her group (see fig. 19.3).

Once the data have been collected, the students in each group should discuss how to organize the bar graph. Since height is involved, it is advised that the bars be vertical. Distribute graph paper and have the students draw a set of axes and label them. They will have to decide whether to record the results in inches, feet, or a combination of both. They should also decide which multiples to use so that everyone's height and distance jumped can fit on the graph properly. Remind the students that equal spacing is necessary between bars and between the numerals listed along the vertical axis. Have them identify a title and insert it on the graph.

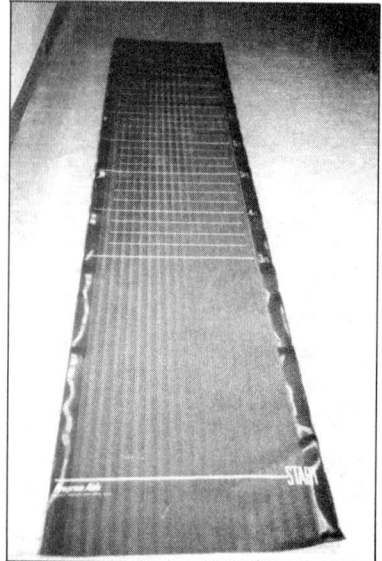

Fig. 19.1

Fig. 19.2

Fig. 19.3

Using the Computer as a Tool

A multiple bar graph option should be available on the software selected. After recording the heights and distances jumped for the members in their groups, the students should work at the computer in pairs to enter the data, construct the graph, and explore other software options (e.g., adjust the scaling and convert to different graph formats). Have them print out their graphs for comparison; the data and the graphs should be stored on a disk.

Questions for Discussion
1. "What is this graph about?" [RD or RBW]
2. "Who is taller, [insert name] or [insert name]?" [RBW]
3. "Who jumped farther, [insert name] or [insert name]?" [RBW]
4. "Who is the tallest?" [RBW]
5. "Who jumped the farthest?" [RBW]
6. "Who is able to jump at least his or her height?" [RBW]
7. "Why do you think that it is a measure of physical fitness to be able to jump your height?" [RBY]
8. "Think of a question you can ask about this graph."

Writing and Reading

After the students have discussed the answers to the questions, ask them to write a description of the graph (i.e., what it means to them). Ask them to share their work by reading their stories aloud or switching papers among groups. Allow group members to compare the graphs with the written verbal descriptions of the graphs. Allow the students to analyze each other's work, encouraging them to question and criticize constructively.

ACTIVITY 20

Topic: Height compared to foot length (or forearm length)

Graph Form: Double bar graph

Graph Title: How Does Your Height Compare with Your Foot Length (or Forearm Length)?

Objectives
1. To measure one's height and foot length (or forearm length) using metric units of measure
2. To collect, organize, and interpret data
3. To construct a double bar graph
4. To compare height and foot length of students arranged in small groups of four or five
5. To use the computer as a tool
6. To answer comprehension questions based on information in a specific graph
7. To express the meaning of a graph in prose

Vocabulary: Graph, double bar graph, height, foot length (or forearm length), metric units of measure, centimeters, meter

Materials: Chalkboard and chalk (or overhead projector, transparencies, and markers); 1/4" graph paper (Appendix 6) or 5-mm graph paper (Appendix 7), pencils, and rulers for each student; metric tape measures attached to wall; recording sheets (similar to Appendix 11); computers and graphing software

Procedure

Ask the students, "How do you think a person's height and foot length are related?" Allow them to discuss what they perceive this relationship to be.

Allow the students to work in groups of four or five to measure each other's height. Attach two metric tape measures end to end to the wall in several places around the classroom (see Activity 15). Show the students how to measure each other's height as described in Activity 19. Each student should record the heights of all the members of his or her group on a data sheet similar to Appendix 11.

Groups of students not at a height station should be measuring their foot length, taking off a shoe and standing on a metric tape measure or ruler. Students should work in pairs so that the foot measure can be read properly. Explain how to place the foot on the tape measure or ruler and how to read it. Each student should record the foot lengths of all the members of his or her group.

Once the data have been collected, the students in each group should discuss how to organize the bar graph. Since height is involved, vertical bars should be used. Distribute graph paper and have the students draw a set of axes and label them. The students should decide which multiples to use so that everyone's height can fit on the graph properly. Remind the students that equal spacing is necessary between bar sets and between numerals listed along the vertical axis. Have them identify a title and insert it on the graph.

Using the Computer as a Tool

See comments for Activity 19.

Questions for Discussion

1. "What is this graph about?" [RD or RBW]
2. "How tall is [insert name]?" [RD]
3. "How long is [insert name]'s foot?" [RD]
4. "Who is the tallest? Who is the shortest?" [RBW]
5. "Whose foot is the longest? Whose foot is the shortest? [RBW]
6. "Who is taller, [insert name A] or [insert name B]? Whose foot is longer, [insert name A] or [insert name B]?" [RBW]
7. "From the data you have collected, what is the relationship between a person's height and his or her foot length? Do you think you have collected enough data to conclude that this is generally true? How can we be sure?" [RBW and RBY]
8. "Think of a question you can ask about this graph."

Writing and Reading

After the students have discussed the answers to the questions, ask them to write a description of the graph (i.e., what it means to them). Ask them to share their work by switching papers, data sheets, and graphs among the groups. Allow students to analyze each other's work, encouraging them to question and criticize constructively.

ACTIVITY 21

Topic: Raisin experiment

Graph Form: a. Bar graph
  b. Line plot
  c. Stem-and-leaf plot
  d. Box plot
  e. Double bar graph
  f. Back-to-back stem-and-leaf plot
  g. Multiple box plot

Graph Title: How Many Raisins Are in a 1/2-oz. Raisin Box?

Objectives
1. To estimate the number of raisins in a 1/2-oz. raisin box
2. To count the raisins in a 1/2-oz. raisin box
3. To collect, organize, and interpret data
4. To categorize estimates and counts according to similar responses
5. To construct a bar graph, line plot, stem-and-leaf plot, box plot, double bar graph, back-to-back stem-and-leaf plot, and multiple box plot
6. To use the computer as a tool
7. To identify and discuss the advantages and disadvantages of using different graph forms to display data
8. To answer comprehension questions based on information in a specific graph
9. To express the meaning of a graph in prose

Vocabulary: Graph, bar graph, double bar graph, line plot, stem-and-leaf plot, back-to-back stem-and-leaf plot, box plot, multiple box plot, estimate, range, number line, median, lower quartile, upper quartile, outlier, mode

Materials: Chalkboard and chalk (or overhead projector, transparencies, and markers); 5-mm graph paper (Appendix 7), lined paper, data recording sheet (Appendix 12), napkins, 1/2-oz. boxes of raisins (all of the same brand), pencils, rulers, two different colors of Post-it notes (3″ × 3″) for each child; computers and graphing software

Procedure

Hold up a 1/2-oz. box of raisins and ask students, "How many raisins do you think are in this box?" Write a few estimates on the chalkboard (or on a transparency).

Distribute boxes of raisins and one color of Post-it notes to all students. Ask them not to open the boxes yet. They must estimate how many raisins are in their boxes and write their estimates (large and dark) on the Post-it notes. If you want to keep a record of individual estimates after the actual counting is done, have the students write their initials in the corner of the notes. Have the students post their estimates on the chalkboard, and discuss with them how to organize the estimates by posting the same estimates in columns. The students should record the data on a sheet similar to Appendix 12. See figure 21.1. (Collect the boxes of raisins if the counting is to be done in another lesson. Be sure to save the Post-it notes of the estimates to compare with actual counts in subsequent lessons.)

a. *Bar graph:* Distribute graph paper. Have the students construct a set of axes and label them. It is suggested that the bars be arranged vertically so that the bar graph can be compared with the line plot in the next section. Discuss the range of the estimates and how to list them along the horizontal axis. Depending on the range of estimates, discuss the possibility of grouping the data (e.g., estimates between

Name of Student	Guess	Actual Count
Chris	25	38
Reagan	20	34
Amie	25	35
Ann	24	40
Jamie	25	32

Fig. 21.1. Students' entries on data sheet

10 and 14; 15 and 19; etc.). Be sure that students identify and insert a graph title. See figure 21.2. Discuss the answers to questions 1–3 and 13 below.

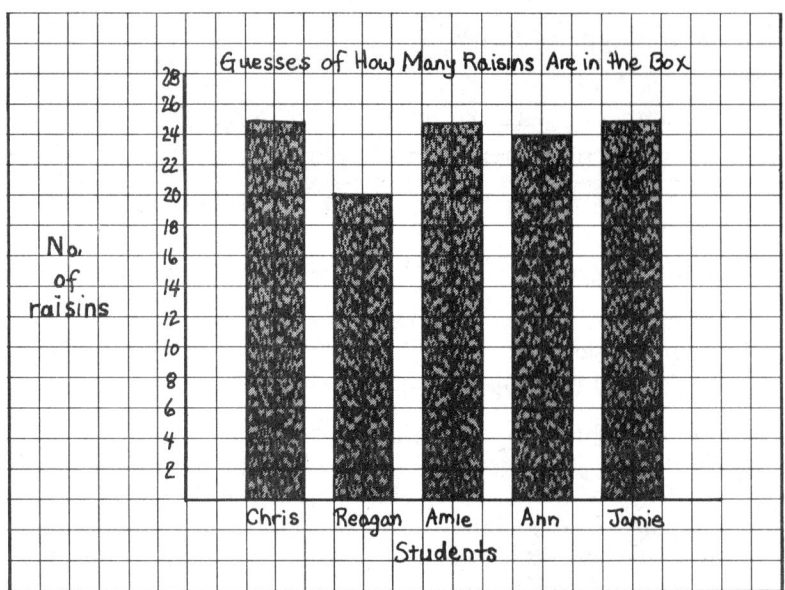

Fig. 21.2. Sixth-grader's graph of estimates

b. *Line plot:* Have the students draw a number line identifying the lowest and highest estimates. Then have them construct a line plot, using the information from the Post-it notes displayed on the chalkboard (or wall or chart). Be sure that the spacing between numerals is equal. See figure 21.3.

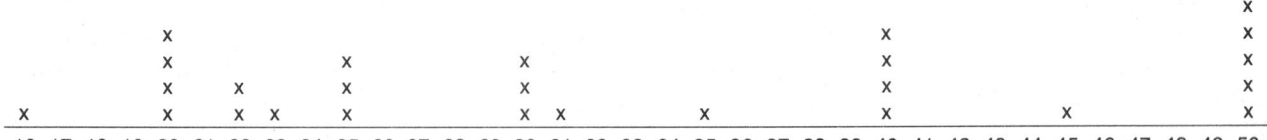

Fig. 21.3. Sixth-graders' example of a line plot for estimates of raisins in a box

Ask the students to compare the information displayed in the line plot with the information displayed in the bar graph (with ungrouped or grouped data). Discuss the answers to questions 1–5 and 13.

c. *Stem-and-leaf plot:* Since most of the estimates will probably be two-digit numbers, have the students rearrange the Post-it notes on the board in rows according to the tens digit (i.e., all single-digit estimates in one row, all estimates in the tens in the second row, etc.). Have students convert this listing to a stem-and-leaf graph. See figure 21.4. Identify the lowest estimate, the highest estimate, the median, and the mode. (Be sure to review the meaning of *median* and *mode*.)

Have the students rotate their stem-and-leaf plots 90 degrees counterclockwise and compare them with

Tens	Ones
1	6
2	0000223555
3	00015
4	00005
5	00000

1 | 6 means 16

Fig. 21.4. Sixth-graders' example of a stem-and-leaf plot for estimates of raisins in a box

the bar graph (ungrouped or group data) and the line plot. Allow the students to discuss the similarities and differences. Discuss the answers to questions 1–7, 12, and 13.

d. *Box plot:* Depending on the students' ability, have them construct a box plot of the estimates by identifying the lowest estimate, the highest estimate, the median, the lower quartile, and the upper quartile. See figure 21.5. Discuss the answers to questions 1–3, 8, 9, 12, and 13.

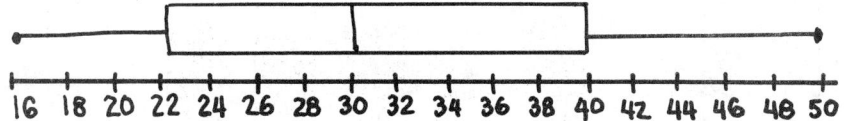

Fig. 21.5. Example of a box plot using data from estimates of raisins in a box

After the students have analyzed the data from the estimates, they are ready to count the number of raisins in their boxes. Distribute napkins, boxes of raisins (if this is a continuation lesson) and a different color of Post-it notes; have the students write their actual count (large and dark) on the notes. Have the students write their initials in the corner of their Post-it note if you want to keep a record of individuals' actual counts to compare with their estimates (see question 3a, below). Have the students arrange the actual counts on the chalkboard similar to the way the estimates were arranged. (If this is a continuation lesson, reattach the Post-it notes of the estimates.) Have the students discuss what they are observing about the actual counts in comparison to the estimates.

Using the data from the actual counts, the students can construct a bar graph, a line plot, a stem-and-leaf plot, and a box plot. Depending on the students' ability, they may go on to comparing the estimates and the actual counts formally by constructing a double bar graph, a back-to-back stem-and-leaf plot, or a multiple box plot.

e. *Double bar graphs:* Have the students group the data from the estimates and group the data from the actual counts so that it will be possible to construct a double bar graph. Allow them to decide whether the bars should be vertical or horizontal. Be sure they label the axes, identify and insert a title, and include a key to represent estimates and actual counts. A sample double bar graph is shown in figure 21.6. Discuss the answers to questions 1, 2a, 3a, 11, and 13.

f. *Back-to-back stem-and-leaf plot:* Using the stem-and-leaf plot constructed for the estimates, allow the students to attach a column to the left to represent the units digits of the actual count (see fig. 21.7). Have them identify the lowest actual count, the highest actual count, and the median. Using this back-to-back plot, ask them to compare the estimates to the actual count. Also, have them compare the

Fig. 21.6. Sixth-grader's double-bar graph

47

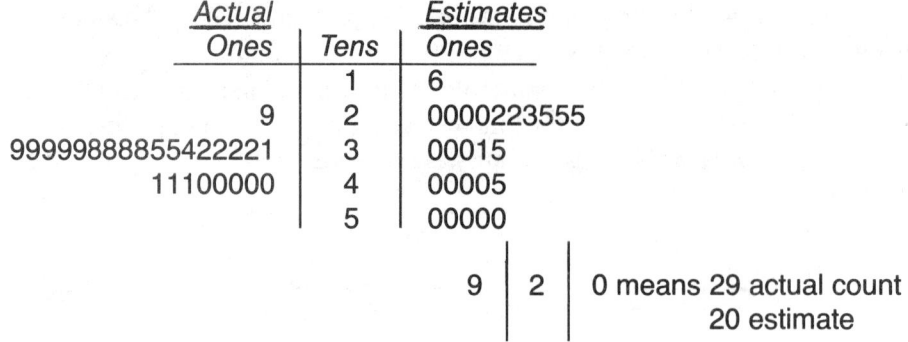

Fig. 21.7. Sample of back-to-back stem-and-leaf plot using data of actual counts and estimates of raisins in a box

information displayed in the back-to-back stem-and-leaf plot with the information compared in the double bar graph. Discuss answers to questions 1, 2a, 3a, and 10–13.

g. *Multiple box plot:* Using one axis to indicate lows, highs, medians, and lower and upper quartiles, have the students construct a box plot for the estimates and the actual counts (see fig. 21.8). Have them compare the results of the estimates with the actual counts. Also, ask them to compare the information displayed in the double bar graph, the back-to-back stem-and-leaf plot, and the multiple box plot. Discuss the answers to questions 1–3, 2a, 3a, 9, and 11–13.

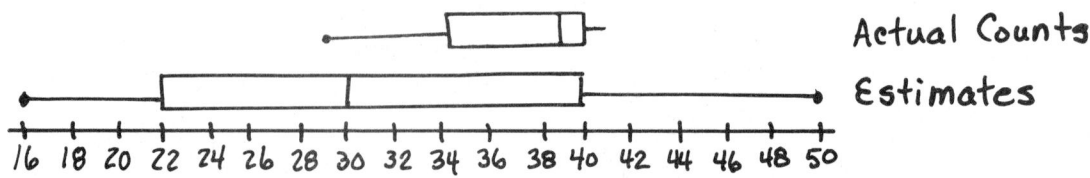

Fig. 21.8. Sample of a multiple-box plot using data of actual counts and estimates of raisins in a box

Using the Computer as a Tool

Graphing software that has the capability of constructing line plots, stem-and-leaf plots, and box plots as well as bar graphs and double bar graphs would be desirable for this activity. However, at the time of this writing, software with all these capabilities was not available for elementary and middle school levels. It is possible to obtain software to construct stem-and-leaf and box plots. With the emphasis being placed on these new statistical techniques, it is likely that more of this type of software will begin to appear.

After the students have had experience constructing the graphs and plots by hand, they should have the opportunity to explore the graphing and plotting options of various pieces of software. Perhaps some groups of students could examine possibilities of graphing the data using bar graphs or double bar graphs, while others use software to apply the plotting techniques. The groups should then exchange the different software to complete their analyses.

The graphs and plots should be printed out to examine and discuss similarities and differences. The data, graphs, and plots should be saved on a disk.

Questions for Discussion

1. "What is this graph or plot about?" [RD or RBW]

2. "Which estimate occurred most frequently [i.e., the mode]? Which estimate occurred the least?" [RBW]

2a. "Which actual count occurred most frequently [i.e., the mode]? Which actual count occurred the least?" [RBW]

3. "Which estimate is the lowest? The highest? What is the range of the estimates?" [RBW]

3a. "Which actual count is the lowest? The highest? What is the range of the actual counts? Whose estimate was the closest to his or her actual count?" [RBW]

4. "Using the line plot or stem-and-leaf plot, how can you find the median? What is the median of the estimates?" [RBY]

5. "What is similar about a bar graph and a line plot? What is different? Which do you prefer? Why?" [RBY]

6. "What is similar about a stem-and-leaf plot and a bar graph? What is different? Which do you prefer? Why?" [RBY]

7. "What is different about a line plot and a stem-and-leaf plot? What is the same? Which do you prefer? Why? [RBY]

8. "What is different about a bar graph and a box plot? Which do you prefer? Why?" [RBY]

9. "What information can you obtain from a stem-and-leaf plot that you cannot obtain from a box plot? When do you think it would be more appropriate to use a box plot than a stem-and-leaf plot?" [RBY]

10. "What is the same [or different] about a double bar graph and a back-to-back stem-and-leaf plot? Which do you prefer? Why?" [RBY]

11. "What brand of raisins did we use? How do you think another brand would compare?" [RBY]

12. "How can we determine which is the best graph form to use to display our data?" [RBY]

13. "Think of a question you can ask about the graphs and plots."

Writing and Reading

After the students have constructed and discussed different graphs and plots, ask them to write a description of the graph or plot (i.e., how they constructed it and what it means to them). Ask them to share their work by switching papers, data sheets, and graphs and plots. Allow the students to analyze each other's work, encouraging them to question and criticize constructively.

Note: This is not meant to be an introduction to line plots, stem-and-leaf plots, or box plots. The concepts involved should be developed prior to doing the respective parts of this activity (see Landwehr and Watkins 1986). Also, all the parts of this activity are not meant to be done in one lesson.

There could be several variations or extensions for this activity. Students could work in small groups and compare estimates and actual counts among themselves using bar and double bar graphs only. Or different brands of 1/2-oz. boxes of raisins could be used. Both estimates and actual counts of each could be compared and discussed, thus raising issues of packaging, weight, size of raisins, and so on.

Ideas for this activity were taken from Helene Silverman (Lesson Demonstration, City College—CUNY, July 1987; 1988). Similar ideas can be found for guessing the colors of M & M's in a certain size of bag and analyzing the actual colors after counting (see Corwin and Friel 1988).

ACTIVITY 22

Topic: Height of a plant over a period of time

Graph Form: Line graph

Graph Title: Height of My Plant from (insert date) to (insert date)

Objectives

1. To plant a bean or seed and record its height over a set period of time
2. To collect, organize, and interpret data
3. To determine the difference between data displayed in a bar graph and data displayed in a line graph
4. To construct a line graph
5. To use the computer as a tool
6. To answer comprehension questions based on information in a specific graph
7. To express the meaning of a graph in prose

Vocabulary: Graph, line graph, height, a period of time, lima bean, seed, plant, soil, millimeters, centimeters

Materials: Chalkboard and chalk (or overhead projector, transparencies, and markers); small water pitchers; adhesive labels, old newspapers, plastic or Styrofoam cups, soil, lima beans or seeds, centimeter rulers, 5-mm graph paper (Appendix 7), data recording sheet (Appendix 13), and pencils for each student; computers and graphing software

Procedure

Ask the students, "If we plant a lima bean, how tall do you think it will grow in three weeks?" Write the students' responses on the chalkboard. "So, all we have to do is plant this bean and it will grow? What does a plant need to grow? [water, light, air, and nutrients from the soil] How must we care for our planted seeds to help them grow into plants?"

Distribute adhesive labels so that the students can write their names on them to label their plants. Then distribute newspapers to protect desks, plastic or Styrofoam cups with soil, and two seeds per student. Have the students insert a finger about 2.5 cm into the soil, place a lima bean or seed into the hole, and cover it. They should leave some space and plant the other seed similarly. After covering the seeds, they should moisten the soil. (Avoid applying too much water, which would drown the seeds.) Place all the cups on a window sill so that they can be exposed to sunlight, and assign students the responsibility for watering the plants daily.

Once the plants begin to break through the soil, have the students measure the height in millimeters and record it on a data collection sheet (Appendix 13).

After collecting data for two or three weeks, the students should construct a line graph. Distribute graph paper and have them draw a set of axes. Indicate the dates (equally spaced) along the horizontal axis and the height (beginning at 0 where the two axes intersect) on the vertical axis. Review the meaning of the lines and boxes. Be sure that the numerals written along the vertical axis are equally spaced. Have the students label the axes, plot the points, connect the points, and insert a title on the graph.

Using the Computer as a Tool

Select graphing software that can be used to construct a line graph as well as other graph forms. Once students are familiar with the software, allow them to work at computers in pairs to enter the data, construct the graphs, and compare the results of displaying the data in different graph forms. Also have them experiment with different scales. The graphs should be printed out to facilitate the discussion of comparisons. The data and the graphs should be saved on a disk.

Questions for Discussion

1. "What is this graph about?" [RD or RBW]
2. "How tall was the plant on [insert a date]?" [RD]
3. "On which date was the plant half its size in comparison to the last date recorded?" [RBW]
4. "Did all the beans grow? Why? Why not?" [RBY]
5. "What were the conditions that contributed to the growth of your plant? Could you have improved any of the conditions so that the plant would have grown faster or taller? How could we investigate this?" [RBY]
6. "Why did we use a line graph and not a bar graph to record the height of our plants?" [RBY]
7. "Think of a question that you can ask about your graph."

Writing and Reading

After the students have discussed the answers to the questions, ask them to write a description of the activity and the graph, telling how they conducted the experiment and what the graph means to them. Writing about the activity can also be in the form of an ongoing log or journal. Ask them to share their work by switching papers, data sheets, and graphs. Allow students to analyze each other's work, encouraging them to question and criticize constructively.

ACTIVITY 23

Topic: Height (or weight) over time (*Note:* For some students, weight may be a sensitive issue. Also, the data may have to be collected for four or five months before the graph can be completed.)

Graph Form: Line graph

Graph Title: How Much I Have Grown (or Gained) from (insert a date) to (insert a date)

Objectives
1. To measure one's height or weight in metric units
2. To collect, organize, and interpret data
3. To construct a line graph
4. To determine the difference between data displayed in a bar graph and data displayed in a line graph
5. To use the computer as a tool
6. To answer comprehension questions based on information in a specific graph
7. To express the meaning of a graph in prose

Vocabulary: Graph, line graph, height, weight, metric units, centimeters, meters (or kilograms)

Materials: Metric tape measures attached to wall (see Activity 15) or a metric scale; data collection sheets (Appendix 13), 5-mm graph paper (Appendix 7), rulers, and pencils for each student; computers and graphing software

Procedure

Ask the students, "How much do you think you will grow during the next four or five months? How can we find out if your prediction comes true?" (Keep a record of height.) *Note:* This could be a continuation of Activity 15, where students were measured for constructing a small-group bar graph.

Allow the students to work in groups of four or five to measure each other's height. Have each student record the date and his or her height on a data collection sheet. This should be done consistently once or twice a month for four or five months (e.g., the first and third Mondays of the month).

Once the data have been collected (i.e., approximately eight measures of height for each individual), discuss how to organize and display the data. Distribute graph paper and have the students draw a set of axes. They should arrange the dates equally spaced along the horizontal axis and list numerals for height (in centimeters) along the vertical axis. Since the vertical axis is to begin with zero, discuss appropriate multiples and spacing of the numerals so that the highest measure will fit on the graph. Be sure to review the meaning of the boxes and lines and how to plot a point located by the intersection of two grid lines. The axes should be labeled and a title inserted on the graph.

Using the Computer as a Tool
See Activity 22.

Questions for Discussion
1. "What is this graph about?" [RD or RBW]
2. "How tall were you on [insert a date]?" [RD]
3. "When were you [insert a measure] centimeters tall?" [RD]
4. "How much did you grow between [insert a date] and [insert a date]?" [RBW]
5. "How old were you at the time of the first measure? At the time of the last measure? Can this information be found directly from the graph? What must be done?" [RBY]
6. "How much do you think you will grow by the time you reach the [insert a grade level] grade?" [RBY]
7. "In Activity 15 we constructed a bar graph to represent heights. In this activity we constructed a line

graph. What is different about the data collected in this activity that requires a different type of graph?" [RBY]

8. "During the time you have been recording your height, how frequently did you have to buy new clothes because you outgrew your old ones, not because they wore out?" [RBY]

9. "Think of a question you can ask about the graph."

Writing and Reading

After the students have discussed the answers to the questions, ask them to write a description about the activity and the graph. Writing about the activity can be in the form of an ongoing log or journal. Ask them to share their work by switching papers, data sheets, and graphs. Allow the students to analyze each other's work, encouraging them to question and criticize constructively.

ACTIVITY 24

Topic: Daily temperature

Graph Form: a. Line graph
b. Multiple line graph

Graph Title: a. What Are Typical Morning (or Afternoon) Temperatures over a Five-Day Period?
b. How Does the Temperature Vary from Morning to Afternoon over the Course of Five Days?

Objectives
1. To determine the outdoor temperature in the morning and the afternoon, by reading a thermometer
2. To collect, organize, and interpret data
3. To construct a line graph and a multiple line graph
4. To determine the difference between data displayed in a line graph and data displayed in a bar graph
5. To compare morning and afternoon temperatures using a multiple line graph
6. To use the computer as a tool
7. To answer comprehension questions based on information in a specific graph
8. To express the meaning of a graph in prose

Vocabulary: Graph, line graph, multiple line graph, key or legend, temperature, thermometer, Celsius, Fahrenheit, degrees

Materials: Overhead projector, transparency of a thermometer scale (either Celsius or Fahrenheit); an outdoor Celsius or Fahrenheit thermometer; a data collection sheet (Appendix 14); 5-mm graph paper (Appendix 7), colored pencils, and pencils for each student; computers and graphing software

Procedure

Ask the students, "How does the outside temperature vary from the morning to the afternoon? From the morning to the evening? How can we be sure?" (Keep a record over the course of several days.)

Distribute data collection sheets to the students. Using an overhead transparency of a thermometer scale (either Celsius or Fahrenheit, to correspond with the scale to be used), review how to read a thermometer. An outdoor thermometer should be attached outside a window so it is clearly visible to the students. For five consecutive days or more, have the students record the temperature at the same time (e.g., 8:45 A.M. and 2:30 P.M.).

a. *Line graph:* Depending on students' ability, graph morning and afternoon temperatures separately, or only collect data for morning or afternoon; adjust the graph title accordingly. Distribute graph paper and have the students construct a set of axes. Along the horizontal axis list the days or dates, equally spaced, that the temperature was taken. Along the vertical axis, list the temperatures. Since zero is indicated where the two axes intersect, any temperature below zero will require the extension of the vertical axis below the horizontal axis (see fig. 24.1). Be sure that the numerals along the vertical axis are equally spaced. The students should label the axes and insert a title. Discuss the answers to questions 1, 2, and 5–7, below.

b. *Multiple line graph:* Distribute graph paper and have the students construct a set of axes. After setting up the axes and labeling them, have the students select two colored pencils—one to represent the morning temperature and one for the afternoon temperature. If colored pencils are unavailable, a dotted or dashed line can represent morning temperature, for example, with a solid line for the afternoon temperature. Be sure to include a key and an appropriate title (see fig. 24.2). Discuss the answers to all the questions below.

Using the Computer as a Tool

Select graphing software that can be used to construct line graphs and multiple line graphs. Once the students are familiar with the software, allow them to work at computers in pairs to enter the data, construct the graphs, explore alternative graph forms, and adjust the scale. Allow sufficient time for students to discuss their results. The graphs should be printed out to facilitate making comparisons. Also, the data and the graphs should be saved on a disk.

Questions for Discussion

1. "What is this graph about?" [RD or RBW]
2. "What was the temperature on [insert a date] in the morning? In the afternoon?" [RD]
3. "On [insert a date], when was the temperature colder, in the morning or the afternoon? Warmer?" [RBW]
4. "Do you notice any temperature patterns from the morning to the afternoon during the five days? If

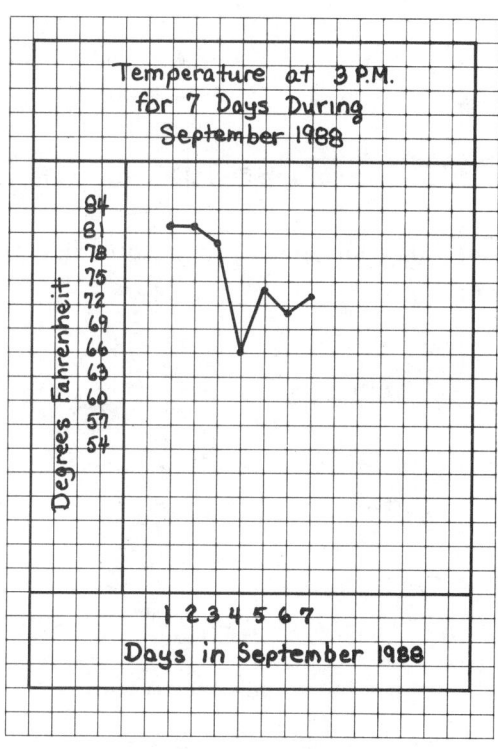

Fig. 24.1. Line graph (Source: The National Weather Service)

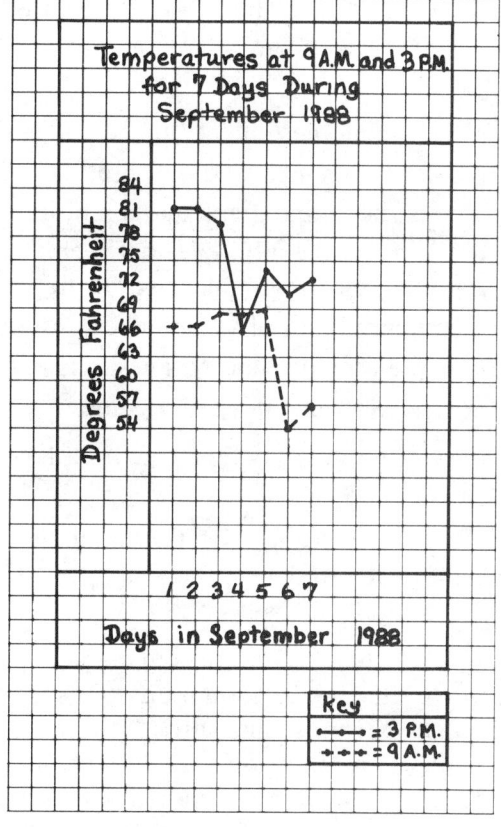

Fig. 24.2. Multiple line graph (Source: The National Weather Service)

so, what do you notice? Do you think this happens most of the time? Why do you think so? How can we be sure?" (Answer: Extend the experiment for five or ten days.) [RBW]

5. "Why do you think people pay attention to the temperature and the weather? Whom do they usually rely on for weather forecasts? Do you think the weather forecaster is always accurate? Why? Why not? About how many times out of ten do you think the weather prediction is correct? How can we be sure? (Answer: Conduct an experiment to compare weather predictions with actual weather over a period of about ten days.) [RBY]

6. "Why were the data collected in this activity displayed in a line graph [or multiple line graph] and not a bar graph [or double bar graph]?" [RBY]

7. "Think of a question you can ask about this graph."

Writing and Reading

After the students have discussed the answers to the questions, ask them to write a description about the activity and the graph. Have them share their work by switching papers, data sheets, and graphs. Allow the students to analyze each other's work, encouraging them to question and criticize constructively.

ACTIVITY 25

Topic: Time of sunrise and sunset

Graph Form: a. Line graph
b. Multiple line graph

Graph Title: a. What Time Does the Sun Rise or Set during a Fourteen-Day Period?
b. How Does the Amount of Daylight Vary over a Fourteen-Day Period?

Objectives
1. To determine the time of sunrise and sunset by checking the times in the newspaper
2. To collect, organize, and interpret data
3. To construct a line graph and a multiple line graph
4. To determine the difference between the data displayed in a line graph and the data displayed in a bar graph, picture graph, or circle graph
5. To compare times of sunrise and sunset over a fourteen-day period using a multiple line graph
6. To answer comprehension questions based on information in a specific graph
7. To express the meaning of a graph in prose

Vocabulary: Graph, line graph, multiple line graph, key or legend, sunrise, sunset, time, the seasons, daylight

Materials: Data collection sheets (Appendix 15), 5-mm graph paper (Appendix 7), pencils, and colored pencils for each student

Procedure

Ask the students, "What time did the sun rise this morning? What time will the sun set this evening? For whom might these be important questions? Why would these be important questions? How can we find out the times of sunrise and sunset?"

Distribute data collection sheets to the students and have them obtain the times of sunrise or sunset over a fourteen-day period by looking in the daily newspaper.

a. *Line graph:* Depending on the student's ability, have them construct a line graph of either sunrise or sunset, or have them make two separate line graphs. Distribute graph paper and have the students construct a set of axes. Along the horizontal axis they should list the dates so that they are equally spaced and along the vertical axis, the times. Since zero is represented at the point of intersection of the two axes, one approach would be to use this point to represent midnight (or the start of the new day). Discuss ways of labeling the times, since the sunsets during a fourteen-day period will have a daily time difference of approximately one minute (see fig. 25.1). Discuss the answers to questions 1–3 and 5.

The times listed along the vertical axis should be equally spaced. Be sure that the axes are labeled and an appropriate title is inserted. Assist students if they need help plotting the points. They may need to be reminded about what the boxes and lines in the graph represent and how to plot a point, which is located by the intersection of two grid lines.

b. *Multiple line graph:* Distribute graph paper and have the students construct a set of axes. Along the horizontal axis list the dates, equally spaced. Since the times of both sunrise and sunset will be listed along the vertical axis, discuss possibilities as to how to do this properly and accurately. Students may have to make a "break" in the graph in order to make all the data fit (see fig. 25.2). Discuss the answers to all the questions below.

Students should select two colors—one to represent times of sunrise, the other for times of sunset. Be sure that a key is included. If colored pencils are unavailable, use a solid line for the times of sunrise, for example, and a dotted or dashed line for the times of sunset. Axes should be labeled and an appropriate title inserted.

Questions for Discussion
1. "What is this graph about?" [RD or RBW]
2. "At what time did the sun rise [or set] on [insert date]?" [RD]
3. "On what day did the sun rise [or set] at [insert time]?" [RD]
4. "During what season are we recording the times that the sun rises [or sets]? As the days progress,

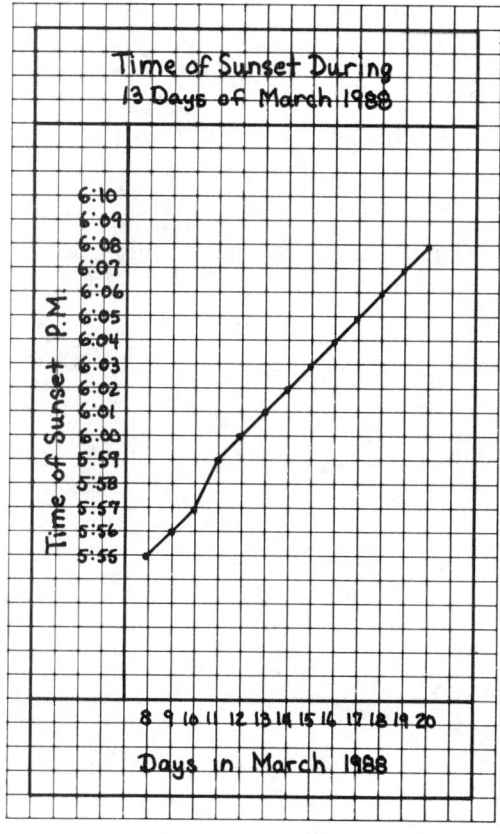

Fig. 25.1. Line graph (Source: *The New York Times*)

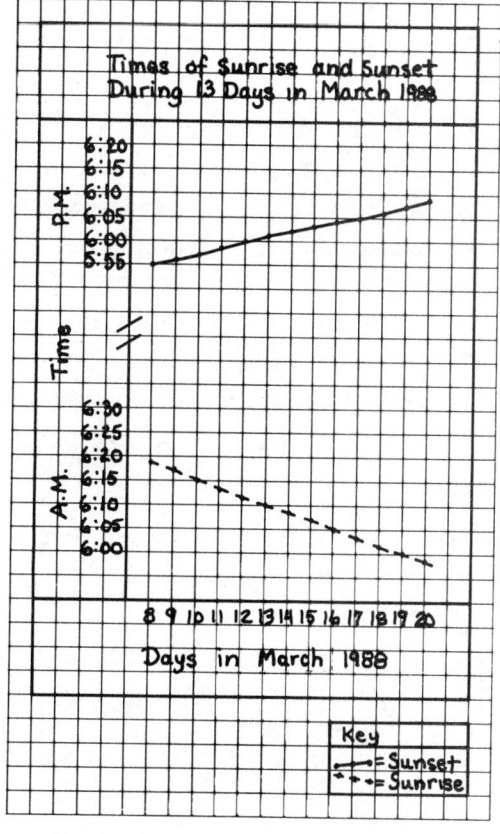

Fig. 25.2. Multiple line graph (Source: *The New York Times*)

what do you notice about the time of sunrise [or sunset]? What is happening to the amount of daylight hours during this season? How would this be the same [or different] during [insert another season]?" [RBY]

5. "What are the characteristics of the data collected in this activity that require them to be displayed in a line graph and not a bar graph, picture graph, or circle graph?" [RBY]

6. "Think of a question you can ask about this graph."

Writing and Reading

After the students have discussed the answers to the questions, ask them to write a description about the activity and the graph. Writing about the activity can be an ongoing log or journal. Ask them to share their work by switching papers, data sheets, and graphs. Allow them to analyze each other's work, encouraging them to question and criticize constructively.

REFERENCES

Arkin, Herbert, and Raymond R. Colton. *Graphs: How to Make and Use Them.* Rev. ed. New York: Harper & Brothers, 1940.

Baratta-Lorton, Mary. *Mathematics Their Way.* Menlo Park, Calif.: Addison-Wesley Publishing Co., 1976.

Bruni, James V., and Helene Silverman. "Graphing as a Communication Skill." *Arithmetic Teacher* 22(1975): 354–66.

Choate, Laura D., and JoAnn K. Okey. "Graphically Speaking: Primary-Level Graphing Experiences." In *Teaching Statistics and Probability,* 1981 Yearbook of the National Council of Teachers of Mathematics, edited by Albert P. Shulte, pp. 33–41. Reston, Va.: The Council, 1981.

Corwin, Rebecca B., and Susan N. Friel. "Statistics: Sampling." *Elementary Mathematician* 2(Spring 1988): Pullout pp. 1–4.

Curcio, Frances R. "Comprehension of Mathematical Relationships Expressed in Graphs." *Journal for Research in Mathematics Education* 18(1987): 382–93.

Huff, Darrell. *How to Lie with Statistics.* New York: W. W. Norton, 1954.

Juraschek, William A., and Nancy S. Angle. "Experiential Statistics and Probability for Elementary Teachers." In *Teaching Statistics and Probability,* 1981 Yearbook of the National Council of Teachers of Mathematics, edited by Albert P. Shulte, pp. 8–18. Reston, Va.: The Council, 1981.

Kirk, Sandra, Paul D. Eggen, and Donald P. Kauchak. "Generalizing from Graphs: Developing a Basic Skill through Improved Teaching Techniques." Paper presented at the annual meeting of the International Reading Association, Saint Louis, May 1980.

Landwehr, James M., and Ann E. Watkins. *Exploring Data.* Quantitative Literacy Series. Palo Alto, Calif.: Dale Seymour Publications, 1986.

Mathis, Judi. "Software Reviews—Turning Data into Pictures: Part 1." *Computing Teacher* 16(October 1988a): 40–48.

———. "Turning Data into Pictures: Part 2." *Computing Teacher* 16(November 1988b): 7–8, 10.

Minnesota Educational Computing Consortium. *MECC Graph,* User's Guide. Saint Paul, Minn.: MECC, 1985.

National Council of Teachers of Mathematics. *Curriculum and Evaluation Standards for School Mathematics.* Reston, Va.: The Council, 1989.

Newman, Claire M., and Susan B. Turkel. "The Class Survey: A Problem-solving Activity." *Arithmetic Teacher* 33(May 1985): 10–12.

Nuffield Foundation. *Pictorial Representation.* New York: John Wiley & Sons, 1967.

Pearson, P. David, and Dale D. Johnson. *Teaching Reading Comprehension.* New York: Holt, Rinehart & Winston, 1978.

Russell, Susan Jo. "Who Found the Most Shells? (Who Cares?)" *Elementary Mathematician* (1988): 4, 9.

Silverman, Helene. Lesson demonstration at the City College of the City University of New York, July 1987.

———. "Big Ideas from Simple Materials: Playing with Data." *New York State Mathematics Teachers' Journal* 38(1988): 48–53.

Stone, Antonia. "When Is a Graph Worth Ten Thousand Words?" *Hands-On!* 11(Spring 1988): 16–18. (Published by TERC, 1969 Massachusetts Ave., Cambridge, MA 02138.)

Tukey, John. *Exploratory Data Analysis.* Reading, Mass.: Addison-Wesley Publishing Co., 1977.

SOFTWARE

Bannasch, Stephen. *Exploring Tables and Graphs.* Middletown, Conn.: Weekly Reader Family Software, 1984.

Edusystems, Inc., and Minnesota Educational Computing Consortium. *MECC Graph.* Saint Paul, Minn.: MECC, 1985.

Lewis, Elaine, et al. *Easy Graph II.* Hanover, N.H.: Houghton Mifflin, 1987.

Stanek, Kevin, and Monte Johnson. *The Quantitative Literacy Project.* Saint Cloud, Minn.: Saint Cloud State University, n.d.

SELECTED BIBLIOGRAPHY

Assad, Saleh. "From Graph to Formula." *Mathematics Teacher* 64(March 1971): 231–32.

Burns, Marilyn. "Alphabet Math." *Instructor* 97(September 1987): 48–50.

Christopher, Leonora. "Graphs Can Jazz Up the Mathematics Curriculum." *Arithmetic Teacher* 30(September 1982): 28–30.

Collis, Betty. "Learning to Like Social Studies." *Computing Teacher* 15(April 1988): 30–33.

Curcio, Frances R. "The Effect of Prior Knowledge, Reading and Mathematics Achievement, and Sex on Comprehending Mathematical Relationships Expressed in Graphs" (Doctoral dissertation, New York University, 1981). *Dissertation Abstracts International* 42(1981a): 3047–3048A.

———. *The Effect of Prior Knowledge, Reading and Mathematics Achievement, and Sex on Comprehending Mathematical Relationships Expressed in Graphs.* Final report. Brooklyn, N.Y.: Saint Francis College, 1981b. (ERIC Document Reproduction Service No. ED 210 185)

———. "Incorporating Graphing and Statistics in the Elementary and Middle School Mathematics Curricula." In *ICME-6 Monograph—Theme Group 7,* edited by John Malone. Perth, Western Australia: Curtin University of Technology, forthcoming.

Curcio, Frances R., and M. Trika Smith-Burke. *Processing Information in Graphical Form.* Paper presented at the annual meeting of the American Educational Research Association, New York, March 1982. Brooklyn, N.Y.: Saint Francis College, 1982. (ERIC Document Reproduction Service No. ED 215 874)

Dickinson, J. Craig. "Gather, Organize, Display: Mathematics for the Information Society." *Arithmetic Teacher* 34(December 1986): 12–15.

Eagle, Edwin. "Toward Better Graphs." *Mathematics Teacher* 35(March 1942): 127–31.

Eng, Kristine L. "Real Graphs, Real Fun, Real Learning." *Learning88* 17(October 1988): 58–62.

Fry, Edward B. *Graphical Comprehension: How to Read and Make Graphs.* Providence, R. I.: Jamestown Publishers, 1981.

Hannah, Larry. "The Data Base: Getting to Know You." *Computing Teacher* 15(August/September 1987): 17–18, 41.

Horak, Virginia M., and Willis J. Horak. "Let's Do It: Collecting and Displaying the Data around Us." *Arithmetic Teacher* 30(September 1982): 16–20.

Johnson, Eleanor M., ed. *Introducing Table and Graph Skills, Book A.* Columbus, Ohio: Xerox Education Publications, 1973–74.

Kelly, Margaret. "Elementary School Activity: Graphing the Stock Market." *Arithmetic Teacher* 33(March 1986): 17–20.

Knowler, Kathleen A., and Lloyd A. Knowler. "Using Teaching Devices for Statistics and Probability with Primary Children." In *Teaching Statistics and Probability,* 1981 Yearbook of the National Council of Teachers of Mathematics, edited by Albert P. Shulte, pp. 41–44. Reston, Va.: The Council, 1981.

Landwehr, James M., Jim Swift, and Ann E. Watkins. *Exploring Surveys and Information from Samples.* Quantitative Literacy Series. Palo Alto, Calif.: Dale Seymour Publications, 1987.

Lynch, Patricia. *Setting Down the Information: Tables, Charts, and Graphs, Book B.* New York: Scholastic, 1977.

MacDonald-Ross, Michael. "How Numbers Are Shown." *AV Communication Review* 25(Winter 1977): 359–409.

Meadows, George C. "Let's Modernize Graph Teaching." *Arithmetic Teacher* 10(May 1963): 286–87.

Murphy, Elaine C. *Developing Skills with Tables and Graphs, Book B.* Palo Alto, Calif.: Dale Seymour Publications, 1981.

National Council of Teachers of Mathematics. *Organizing Data and Dealing with Uncertainty.* Reston, Va.: The Council, 1979.

Nelson, L. Doyal, and Joan Kirkpatrick. "Problem Solving." In *Mathematics Learning in Early Childhood,* Thirty-seventh Yearbook of the National Council of Teachers of Mathematics, edited by Joseph N. Payne, pp. 69–93. Reston, Va.: The Council, 1975.

New York State Education Department. *Graphs and Statistics: A Resource Handbook.* Albany, N.Y.: The University of the State of New York, 1977.

Nibbelink, William. "Graphing for Any Grade." *Arithmetic Teacher* 30(November 1982): 28–31.

Paine, Carolyn. "Graphing Matters." *Learning* 11(January 1983): 38–40.

Russell, Susan Jo, and Susan N. Friel. "Collecting and Analyzing Real Data in the Elementary School Classroom." In *New Directions for Elementary School Mathematics,* 1989 Yearbook of the National Council of Teachers of Mathematics, edited by Paul R. Trafton, pp. 134–48. Reston, Va.: The Council, 1989.

Shaw, Jean M. "Let's Do It: Making Graphs." *Arithmetic Teacher* 31(January 1984): 7–11.

Slaughter, Judith P. "The Graph Examined." *Arithmetic Teacher* 30(March 1983): 41–45.

Smith, Robert F. "Bar Graphs for Five-Year-Olds." *Arithmetic Teacher* 27(October 1979): 38–41.

Strickland, Ruth G. *A Study of the Possibility of Graphs as a Means of Instruction in the First Four Grades of the Elementary Schools.* 1938. Reprint. New York: AMS Press, 1972.

Sullivan, Delia, and Mary Ann O'Neil. "This Is Us! Great Graphs for Kids." *Arithmetic Teacher* 28(September 1980): 14–18.

Trafton, Paul. "The Curriculum." In *Mathematics Learning in Early Childhood,* Thirty-seventh Yearbook of the National Council of Teachers of Mathematics, edited by Joseph N. Payne, pp. 15–41. Reston, Va.: The Council, 1975.

Tufte, Edward R. *The Visual Display of Quantitative Information.* Cheshire, Conn.: Graphics Press, 1983.

van Engen, Henry, and Douglas Grouws. "Relations, Number Sentences, and Other Topics." In *Mathematics Learning in Early Childhood,* Thirty-seventh Yearbook of the National Council of Teachers of Mathematics, edited by Joseph N. Payne, pp. 251–71. Reston, Va.: The Council, 1975.

Woodward, Ernest, and Frances Byrd. "Make Up a Story to Explain the Graph." *Mathematics Teacher* 77(January 1984): 32–34.

APPENDIX 1

A List of Graph Topics Appropriate for Different Grade Levels[1]

Topic and Question	K	1	2	3	4	5	6	7	8
Favorites: What/Who is your favorite . . .									
color?	x	x	x	x					
dinosaur?	x	x	x	x					
fairy-tale character?	x	x	x	x					
toy?	x	x	x	x	x				
type of pet?	x	x	x	x	x				
zoo animal?	x	x	x	x					
holiday?	x	x	x	x	x	x	x	x	x
ice cream flavor?	x	x	x	x	x	x	x	x	x
candy?	x	x	x	x	x	x	x	x	x
snack?	x	x	x	x	x	x	x	x	x
dessert?	x	x	x	x	x	x	x	x	x
day of the week?		x	x	x					
chore (at home)?		x	x						
song?		x	x	x	x	x	x	x	
(nonboard) game?		x	x	x	x	x	x	x	
type of food?		x	x	x	x	x	x	x	
vegetable?		x	x	x	x	x			
fruit?		x	x	x	x	x			
TV show?		x	x	x	x	x	x	x	
classroom job?		x	x	x	x	x			
season of the year?		x	x	x	x	x	x	x	
subject in school?			x	x	x	x	x	x	
fast-food restaurant?			x	x	x	x	x	x	
sport to play?				x	x	x	x	x	
board game?				x	x	x	x	x	
book (title)?				x	x	x	x	x	
story (title)?				x	x	x	x	x	
(least favorite) subject (in school)?				x	x	x	x	x	
pizza topping?				x	x	x	x	x	
school activity?				x	x	x	x	x	
flower?				x	x	x	x	x	
video game?				x	x	x	x	x	
soft drink?				x	x	x	x	x	
weekend activity?					x	x	x	x	
sport to watch?					x	x	x	x	
book type?					x	x	x	x	
movie (title)?					x	x	x	x	

Topic and Question	K	1	2	3	4	5	6	7	8
Favorites (Continued): What/Who is your favorite . . .									
hobby?						x	x	x	x
lucky number?						x	x	x	x
type of apple? (e.g., raw, sauce, juice)						x	x	x	x
insect?						x	x	x	x
rock group?							x	x	x
male singer?							x	x	x
female singer?							x	x	x
male athlete?							x	x	x
female athlete?							x	x	x
athletic shoe brand?							x	x	x
musical instrument?							x	x	x
TV soap opera?							x	x	x
car?							x	x	x
computer?							x	x	x
brand of toothpaste?							x	x	x
author?							x	x	x
radio station?							x	x	x
designer label?								x	x
actor?								x	x
actress?								x	x
Counting, Quantity: How many . . .									
teeth have you lost?	x	x	x						
children drink milk for lunch?	x	x	x	x					
children are in your family?	x	x	x	x	x	x			
children in our class have the same name?		x	x	x					
children did not finish their milk at lunch?		x	x	x					
TVs do you have in your house?		x	x	x	x	x			
letters in your first (or last) name?			x	x					
children are present (or absent) today?			x	x					
vowels (or consonants) in your first (or last) name?			x	x					
coins of each type do we have?			x	x					
children wear (or don't wear) glasses?			x	x	x	x			
sunny days did we have in each week (for one month)?				x	x	x	x	x	
children celebrate a birthday during each month?				x	x	x	x	x	
children celebrate a birthday during each season?					x	x	x	x	
children are right- (or left-) handed?					x	x	x		
windows do you have in your house?					x	x	x		
doors does your house have?					x	x	x		
mirrors do you have in your house?					x	x	x		

Topic and Question	Grade Level
	K 1 2 3 4 5 6 7 8

Counting, Quantity (Continued): How many ...

	K	1	2	3	4	5	6	7	8
rooms does your house have?				x	x				
books have you read in one week (or month)?					x	x	x	x	x
times are the letters of the alphabet used in a 100-word passage?						x	x	x	x
parts of speech (noun, verb, adjective, adverb, other) are in a "random" sentence?						x	x	x	x
riders are there in cars passing the school between 10:00 and 11:00 a.m.?						x	x	x	x
children did each U.S. president have?						x	x	x	x
times does a vowel appear in one newspaper article?						x	x	x	x
vowels occur in 10 lines of prose?						x	x	x	x
kilowatt hours of energy (i.e., electricity) do we use each month?							x	x	x
calories in (some) milk products?							x	x	x
calories in seven favorite foods?							x	x	x

Categorizing, Measurement: What is/are ...

	K	1	2	3	4	5	6	7	8
your bedtime?	x	x	x						
your hair color?	x	x	x	x	x				
your eye color?	x	x	x	x	x				
the types of food we eat for lunch?			x	x	x	x	x	x	
your shoe size?				x	x	x	x	x	
the indoor/outdoor temperature at noon for five days?				x	x	x	x	x	x
type of footwear do you have on?				x	x	x	x	x	x
the national average height/weight of children ages 7-8-9?					x	x	x	x	x
the kinds of cars your parents drive/own?					x	x	x	x	x
size of books we have on our desks?					x	x	x		
the distance you can throw a softball, basketball, etc.?						x	x	x	x
the length of our hands compared to the length of our feet?						x	x	x	x
the height of the tallest child in each of the _____(th) grades?						x	x	x	x
maximum/minimum temperatures for five consecutive days?						x	x	x	x
the population of (selected) countries?						x	x	x	x
the population of our neighboring communities?						x	x	x	x
the length of the five longest rivers in the world?						x	x	x	x
the life span of (selected) animals?						x	x	x	x
our community/city population (over a given period of time)?						x	x	x	x
the number of sit-ups you can do in one minute?						x	x	x	x
the distance you can run in one minute?						x	x	x	x
the approximate area of your (dominant) hand?							x	x	x
the length of our feet compared to our heights?							x	x	x
the stopping distance (in feet) for various car speeds?							x	x	x
the cost of (selected) kinds of cars?							x	x	x

Topic and Question	K	1	2	3	4	5	6	7	8
Categorizing, Measurement (Continued): What is/are ...									
the approximate miles/gallon for (selected) kinds of cars?							x	x	x
the average time of sunrise/sunset for each month of the year?							x	x	x
the approximate land areas of (selected) countries?							x	x	x
the population of (selected) cities?							x	x	x
the height of (selected) mountains?							x	x	x
the barometric pressure recorded at different times during the day?							x	x	x
the depths of the oceans of the world?							x	x	x
the length of the sun's shadow at different times during the day (during different seasons)?							x	x	x
the number of centimeters a candle will burn in a given amount of time?							x	x	x
the frequency of the letters used in the English language?								x	x
American casualties resulting from (selected) wars?								x	x
Miscellaneous, Measurement, Categorizing									
How do we travel to school?	x	x	x	x	x	x			
What type of pet do you own?	x	x	x	x	x	x	x		
How tall are you? (over a period of time)		x	x	x	x	x	x	x	
How tall is each child in our class?			x	x	x	x			
How much does each child weigh?			x	x	x	x			
In what type of dwelling do you live?			x	x	x	x			
How do you spend each hour of a school day?						x	x	x	x
How do you spend each hour on a Saturday?						x	x	x	x
How do you spend your daily school allowance?						x	x	x	x
How much milk does each class consume in one day?						x	x	x	
How does your height/weight change over a period of 4 months?							x	x	x
How fast does your plant grow (result of seed planting)?							x	x	x
What is the relationship between height and weight?							x	x	x
How does the distance we can jump (broad/long jump) compare with our height?							x	x	x
How much water do we consume in one month?							x	x	x
What were the European countries represented by explorers who traveled to the New World between 1492 and 1693?							x	x	x
What was our city's public school enrollment from 1900 to 1980?							x	x	x
How do (selected) stock prices change over a period of time?								x	x

1. The graph topics were collected from various sources (see Nuffield 1967; Shaw 1984). The topics were then rated by twenty-four elementary and middle school teachers, with each having over five years of teaching experience at multiple grade levels. Grade levels were agreed on by at least 75 percent of the teachers.

APPENDIX 2

How to Make Reusable Teaching Aids

Floor Grid (for people graphs in Activities 1 and 2—see fig. A2.1)

Materials: heavy brown packing paper, oilcloth, or heavy plastic; yardstick; 1" or 3/4" masking tape or cloth tape; pencil, marker, or paint

Directions

a. Measure a 6'6" × 8'8" piece of paper, oilcloth, or plastic. If the paper is only 3' wide, you will have to attach two 3' widths together, or just make a narrower floor grid. If you use a smaller width, the choices will be limited to approximately three categories.

b. Mark off 1' × 1' squares using 1" masking tape, cloth tape, or markers to make the horizontal and vertical lines.

c. Depending on the quality and durability of the material used, you may have to cover the floor grid with clear contact paper.

Object Grid (for object and picture graphs in Activity 4—see fig. A2.2)

Materials: 22" × 28" piece of white oaktag or tagboard, clear contact paper, ruler, pencil, permanent black marker

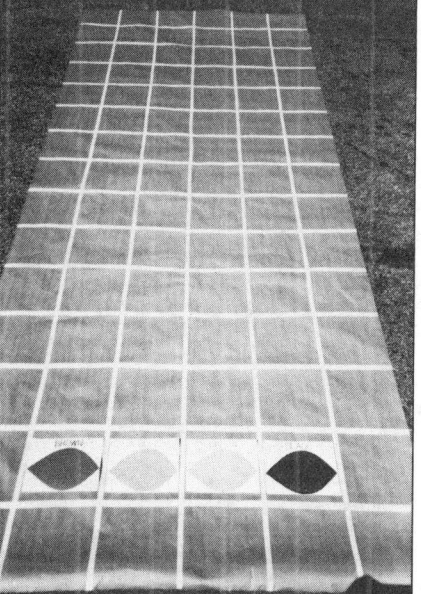

Fig. A2.1

Directions

a. Measure 4" × 4" squares with a pencil.

b. Draw over the horizontal and vertical lines with a permanent black marker.

c. Cover the grid with clear contact paper to protect the surface.

d. To use the grid for a picture graph, cut a small 1/2" slit in the middle of the top edge of each square. A paper clip should be inserted to hold small pictures or drawings for the picture graph.

From Choate & Okey (1981, pp. 34–37).

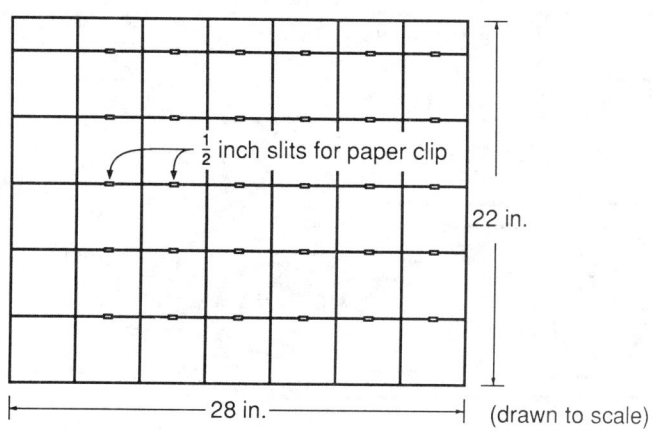

Fig. A2.2

APPENDIX 3

Picture Labels for Object Graph and Picture Graph

APPENDIX 4 **Large-Box Graph Paper**

APPENDIX 5 1-cm Graph Paper

APPENDIX 6   1/4" **Graph Paper**

APPENDIX 7 5-mm Graph Paper

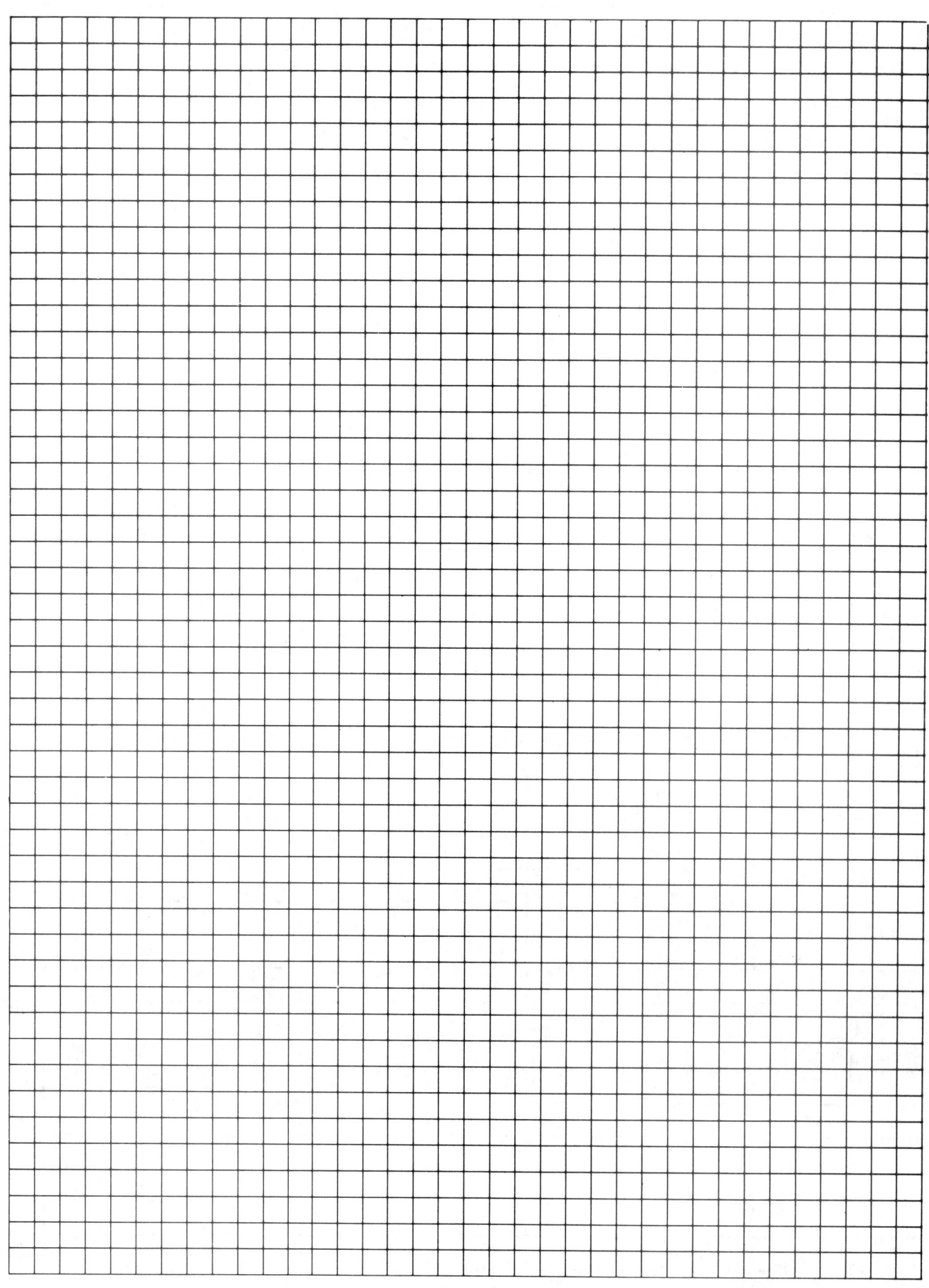

APPENDIX 8

Daily Activities Data Collection Sheet

Name _____   Date _____

School _____   Class _____

Due _____

1. Fill in each hour time slot with an activity. Think of a typical day when filling in the activity.

    A.M.   P.M.
    12:00 _____   12:00 _____
    1:00 _____   1:00 _____
    2:00 _____   2:00 _____
    3:00 _____   3:00 _____
    4:00 _____   4:00 _____
    5:00 _____   5:00 _____
    6:00 _____   6:00 _____
    7:00 _____   7:00 _____
    8:00 _____   8:00 _____
    9:00 _____   9:00 _____
    10:00 _____   10:00 _____
    11:00 _____   11:00 _____

2. List each activity and the total number of hours spent engaged in the activity.

    Activity   *Number of Hours*
    _____   _____
    _____   _____
    _____   _____
    _____   _____
    _____   _____
    _____   _____
    _____   _____
    _____   _____
    _____   _____
    _____   _____

(Continued on next page.)

Name _____

Date _____

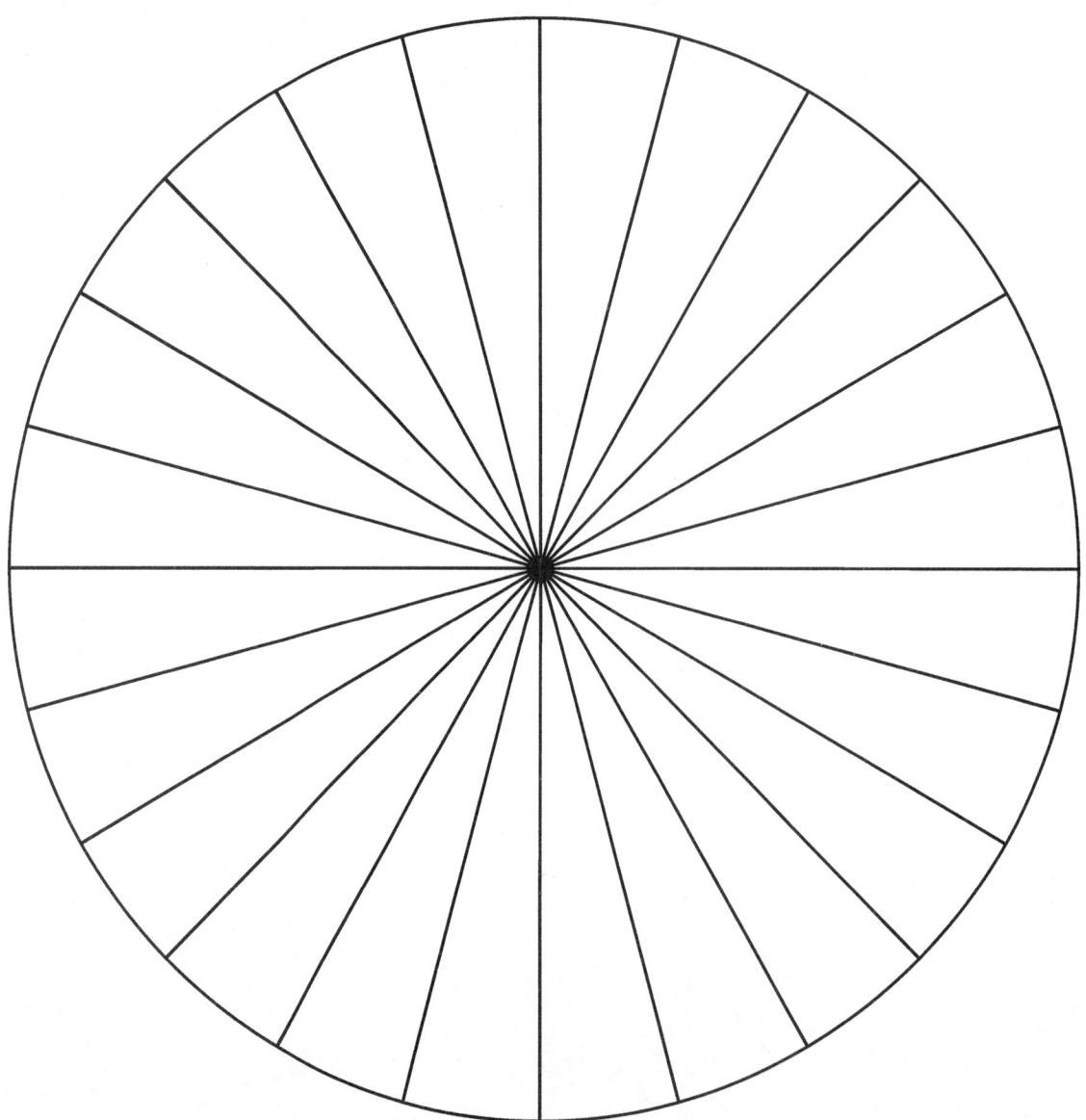

3. Using each space to represent one hour, label each activity in the graph. Use a different color to highlight each activity.

   a. How many hours do you spend sleeping? Eating? Watching TV?
   b. For which activity do you spend the most amount of time?
   c. For which activity do you spend the least amount of time?
   d. After analyzing how you spend your time on Saturday, is there anything you would like to change? If so, what and how?
   e. How would a weekday graph be different from this one?
   f. How would a summer day graph be the same/different as compared to this graph?
   g. Think of a question that you can ask.

APPENDIX 9

Supplemental Graph Reading Activities and Answer Key

All questions for Graphs 1–7 are arranged according to the following levels of comprehension: questions 1–2, Reading the Data; 3–4, Reading between the Data; 5–6, Reading beyond the Data. Answers are given on page 78.

Graph 1

How John Spends His Daily School Allowance

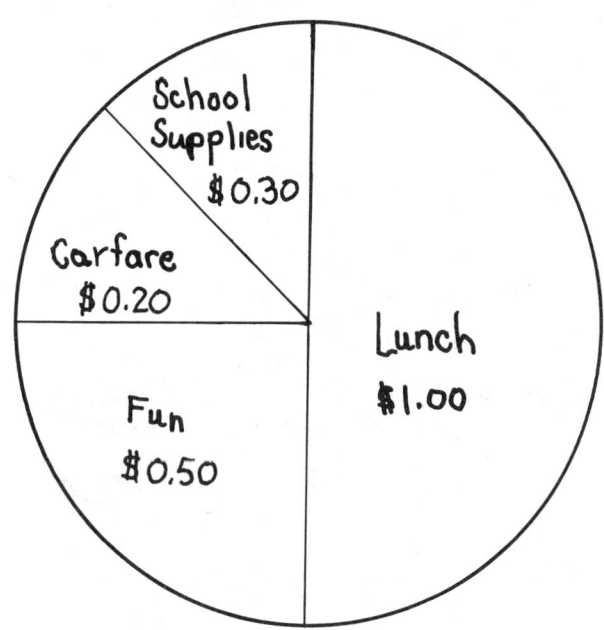

Use the graph above to answer the following questions:

1. What does this graph tell you? 1. _____
   a. The way John spends his money on weekends
   b. The way John spends his money during one complete week
   c. The way John spends his money for vacations
   d. The way John spends his money for one school day

2. How much does John spend on school supplies for one day? 2. _____
   a. $0.20   c. $0.50
   b. $0.30   d. $1.00

3. What is the total of John's daily school allowance? 3. _____
   a. $1.90   c. $3.00
   b. $2.00   d. $4.00

4. How much more does John spend on lunch than on carfare? 4. _____
   a. $0.50   c. $0.80
   b. $0.70   d. $1.20

5. How much money does John need to pay for lunch for five school days? 5. _____
   a. $1.00   c. $4.00
   b. $2.00   d. $5.00

6. What is the total amount of money John needs for five school days? 6. _____
   a. $2.00   c. $10.00
   b. $7.00   d. $14.00

71

Graph 2

How Terry Spends a School Day

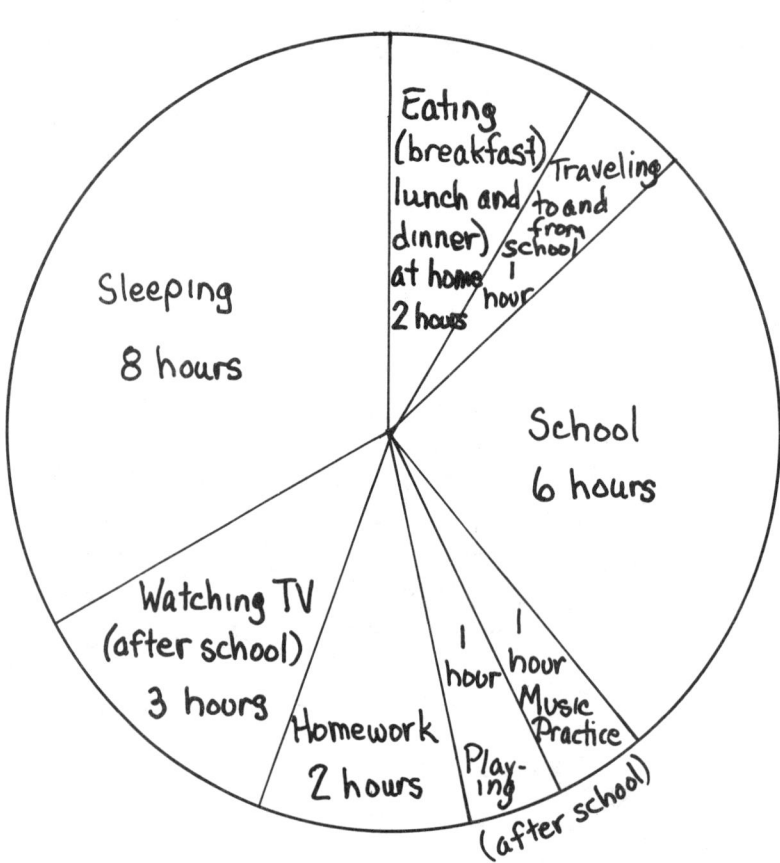

Use the graph above to answer the following questions:

1. How many hours in one day does Terry spend in school?   1. ____
   a. 2 hours   c. 8 hours
   b. 6 hours   d. 14 hours

2. For which of the following does Terry spend three hours a day?   2. ____
   a. Playing after school
   b. Eating
   c. Traveling to and from school
   d. Watching TV

3. What does this graph tell you?   3. ____
   a. Terry spends the greatest amount of time sleeping
   b. Terry spends the least amount of time in school
   c. Terry spends more time playing than watching TV
   d. Terry spends less time sleeping than in school

4. What fractional part of a day does Terry spend in school?   4. ____
   a. $\frac{1}{12}$ b. $\frac{1}{4}$ c. $\frac{1}{3}$ d. $\frac{1}{2}$

5. How many hours a week (not including Saturday and Sunday) does Terry spend on homework?   5. ____
   a. 2 hours   c. 10 hours
   b. 8 hours   d. 14 hours

6. What fractional part of a week (not including Saturday and Sunday) does Terry spend sleeping?   6. ____
   a. $\frac{1}{21}$ b. $\frac{1}{12}$ c. $\frac{1}{5}$ d. $\frac{1}{3}$

Graph 3

Height of the Rodriguez Children in March 1989

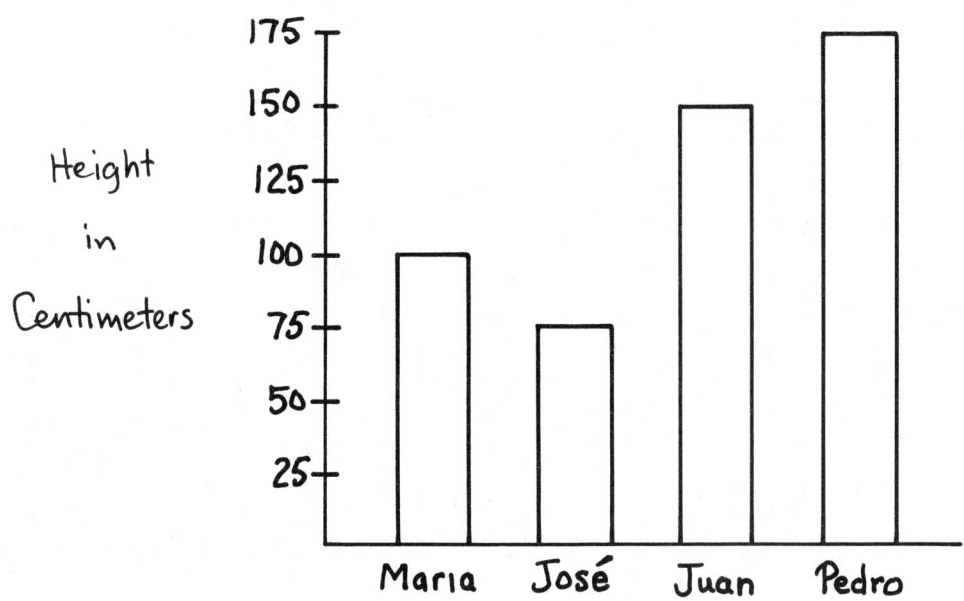

Use the graph above to answer the following questions:

1. What does this graph tell you? 1. ____
   a. The weight of the four Rodriguez children in March 1989
   b. The grades of the four Rodriguez children in March 1989
   c. The height of the four Rodriguez children in March 1989
   d. The age of the four Rodriguez children in March 1989

2. How tall was Maria? 2. ____
   a. 75 inches
   b. 100 inches
   c. 100 centimeters
   d. 125 centimeters

3. Who was the tallest? 3. ____
   a. Juan
   b. Pedro
   c. José
   d. Maria

4. How much taller was Juan than José? 4. ____
   a. 25 centimeters
   b. 50 centimeters
   c. 75 inches
   d. 75 centimeters

5. If Maria grows 5 centimeters and José grows 10 centimeters by September 1990, who will be taller, and by how much? 5. ____
   a. Maria will be taller by 20 centimeters
   b. José will be taller by 20 centimeters
   c. Maria will be taller by 5 centimeters
   d. José will be taller by 5 centimeters

6. If Pedro is 5 years old, which of the following is a correct statement? 6. ____
   a. Pedro is much too short for his age
   b. Pedro could never be that tall for his age
   c. Pedro is of average height for his age
   d. Pedro is thin for his age

Graph 4

The Number of Children in Mr. Kahn's Class Celebrating a Birthday during Each Month of the Year

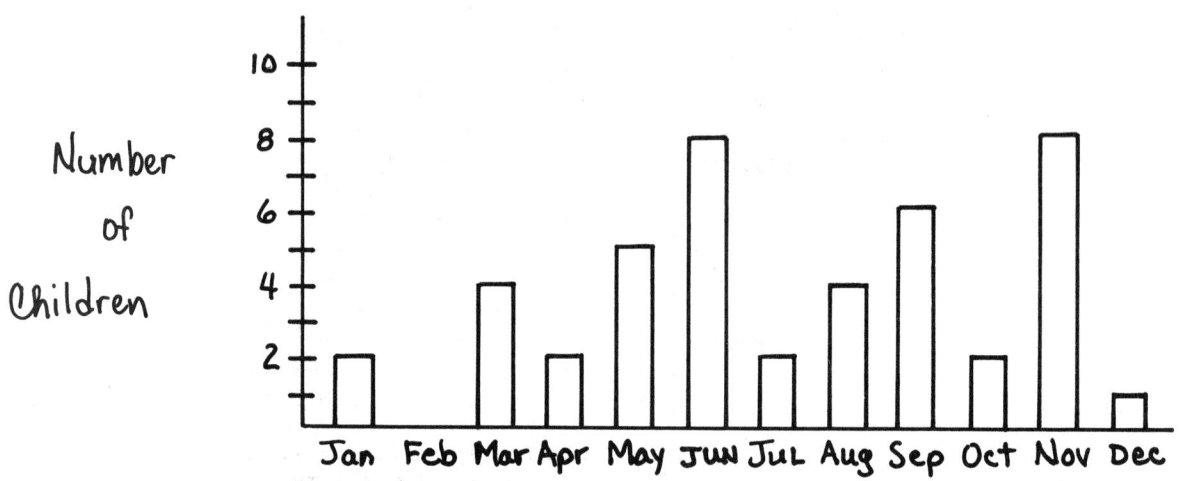

Use the graph above to answer the following questions:

1. How many children celebrate a birthday in February?
   a. 0
   b. 1
   c. 2
   d. 4

   1. ____

2. During which month are there eight children who celebrate a birthday?
   a. May
   b. July
   c. September
   d. November

   2. ____

3. What does this graph tell you?
   a. There are more birthdays during June and November than during any other month of the year
   b. There are more birthdays during May than during any other month of the year
   c. There are fewer birthdays during June and November than during any other month of the year
   d. As the year progresses from January to December, the number of birthdays decreases

   3. ____

4. How many children are in Mr. Kahn's class?
   a. 10
   b. 30
   c. 44
   d. 55

   4. ____

5. What is the probability that the birthday being celebrated in December occurs on 25 December?
   a. $\frac{1}{31}$
   b. $\frac{25}{31}$
   c. $\frac{30}{31}$
   d. 1

   5. ____

6. Sally's birthday is in February. According to the graph, which of the following statements is correct?
   a. Sally was probably born on 29 February
   b. Sally *is not* in Mr. Kahn's class
   c. Sally *is* in Mr. Kahn's class
   d. Sally is the only one celebrating a birthday in February

   6. ____

Graph 5

Average Time of Sunset

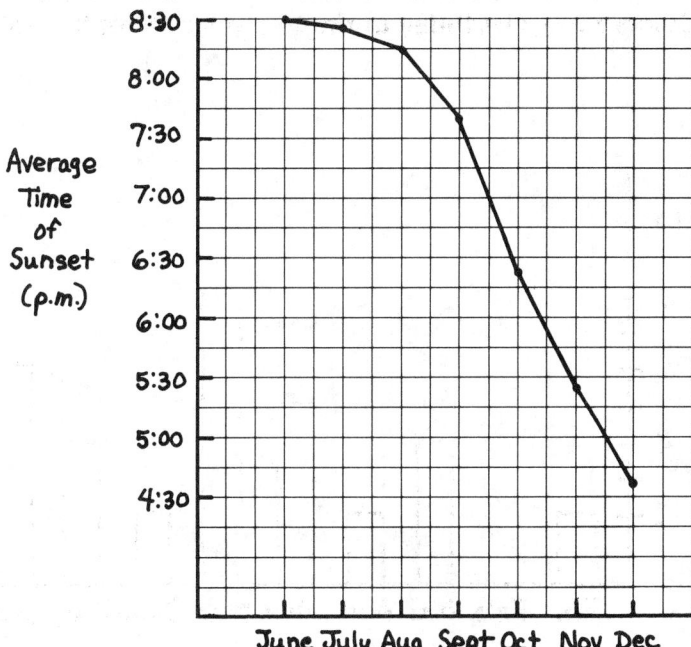

Use the graph above to answer the following questions:

1. What is the average time that the sun sets in October?
   a. 5:00 p.m.   c. 6:15 p.m.
   b. 5:15 p.m.   d. 7:30 p.m.

2. 4:35 p.m. is the average time of sunset during which month?
   a. October
   b. November
   c. December
   d. January

3. As the months progress from June to December, which of the following is true about the average time of sunset?
   a. It gets earlier
   b. It gets later
   c. It remains the same
   d. It first gets earlier and then later

4. How much longer do you have to play outside (before it gets dark) in July than you have in October?
   a. $1\frac{1}{4}$ hours   c. $2\frac{1}{4}$ hours
   b. $1\frac{1}{2}$ hours   d. 3 hours

5. Which of the following graphs represents the average time of sunset from January to June that would make the graph above represent one complete year?

   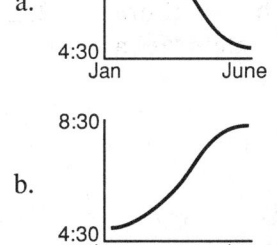
   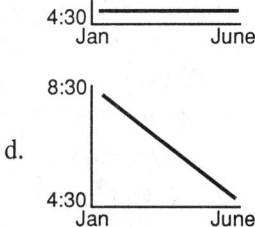

6. As the months progress from June to December, the average time of sunrise gets later. What do you expect to happen to the average number of daylight hours during this time?
   a. Increases
   b. Decreases
   c. Remains the same
   d. First decreases, and then increases

75

Graph 6

The Number of Books the Jones Children Read per Month

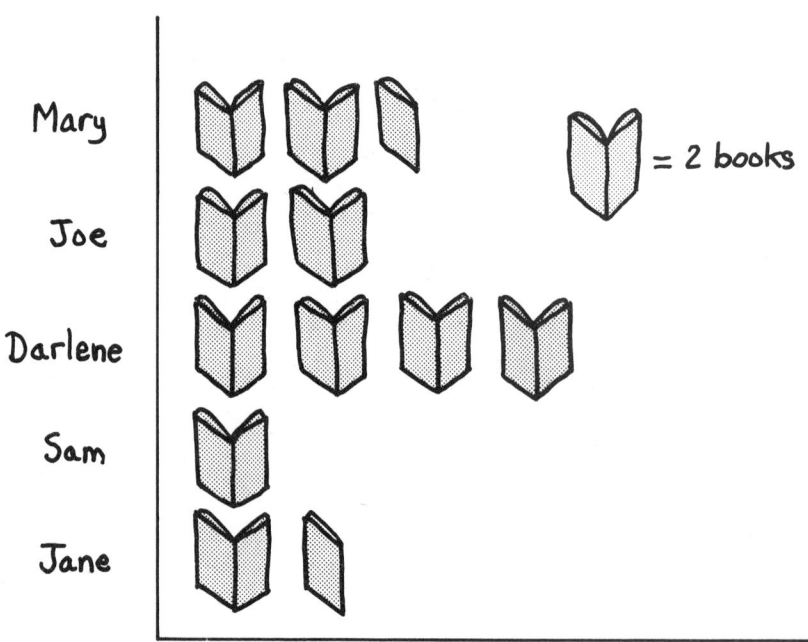

Use the graph above to answer the following questions:

1. How many books did Sam read in one month? 1. ____
   a. 0   c. $1\frac{1}{2}$
   b. 1   d. 2

2. Who read $2\frac{1}{2}$ books in one month? 2. ____
   a. Mary   c. Joe
   b. Jane   d. No one

3. Who read the *least* number of books in one month? 3. ____
   a. Darlene   c. Jane
   b. Sam   d. Joe

4. In one month, how many more books did Darlene read than Jane? 4. ____
   a. $1\frac{1}{2}$ c. 5
   b. $2\frac{1}{2}$ d. $5\frac{1}{2}$

5. At the end of one year, about how many books will Joe have read? 5. ____
   a. 16   c. $30\frac{1}{2}$
   b. 24   d. 48

6. About how many books does Darlene read in one week? 6. ____
   a. 1   c. 4
   b. 2   d. 8

Graph 7

Stamps Collected by Children

Use the graph above to answer the following questions:

1. What does each symbol represent?
   a. One stamp
   b. One-half of a stamp
   c. Four stamps
   d. Twenty-five stamps

2. How many stamps has Tom collected?
   a. $1\frac{1}{2}$ c. $37\frac{1}{2}$
   b. 6   d. 50

3. How many more stamps has Betty collected than Tom?
   a. $1\frac{1}{2}$ c. 6
   b. 3   d. $37\frac{1}{2}$

4. According to the graph, which of the following statements is true?
   a. Jack has $\frac{1}{2}$ more stamps than Betty
   b. Tom has 8 fewer stamps than Jack
   c. Betty has $1\frac{1}{2}$ more stamps than Tom
   d. Betty has 75¢ worth of stamps

5. If the children sold their stamps for the value on the face of the stamp, who, if any of the children, would be able to buy a *new* bicycle using this money?
   a. Tom   c. Jack
   b. Betty   d. None of the children

6. One of Tom's stamps is rare, but he doesn't know it. He and Betty are going to trade one stamp for one stamp. What is the probability that Betty receives Tom's rare stamp?
   a. $\frac{1}{2}$ c. $\frac{1}{6}$
   b. $\frac{1}{12}$ d. $\frac{1}{18}$

Supplemental Graph Reading Activities
Answer Key

Graph 1	Graph 2	Graph 3	Graph 4
1. d	1. b	1. c	1. a
2. b	2. d	2. c	2. d
3. b	3. a	3. b	3. a
4. c	4. b	4. d	4. c
5. d	5. c	5. a	5. a
6. c	6. d	6. b	6. b

Graph 5	Graph 6	Graph 7
1. c	1. d	1. c
2. c	2. d	2. b
3. a	3. b	3. c
4. c	4. c	4. b
5. b	5. d	5. d
6. b	6. b	6. c

APPENDIX 10

Data Collection Sheet

1. For three adults, record their height and the length of their bare right foot to the nearest centimeter.

	#1	#2	#3
Height	____	____	____
Foot Length	____	____	____

2. Record the favorite ice cream flavor of five persons, including yourself. There are two ways; try both. First, simply ask for their favorite flavor.

____	____	____	____	____
#1	#2	#3	#4	#5

   Second, ask for their favorite among chocolate, strawberry, vanilla, and other.

____	____	____	____	____
#1	#2	#3	#4	#5

3. Record the eye color of five persons, including yourself.

____	____	____	____
blue	brown	gray	other

4. Record the hair color of five persons, including yourself.

____	____	____	____	____	____
blond	brown	black	red	gray	other

5. Record the number of letters and spaces in the names of five persons, including yourself. (The name used to sign checks.)

   _____ _____ _____ _____ _____

6. Record the number of each type of coin in your possession right now. (List denomination under each.)

   _____ _____ _____ _____ _____

7. Record the favorite TV program of five persons, including yourself.

   _____ _____ _____ _____ _____

8. Record the birth month of five persons, including yourself.

   _____ _____ _____ _____ _____

Adapted from Juraschek and Angle (1981).

APPENDIX 11

Height Data Collection Sheet

Name _____ Date _____

School _____ Class _____

1. Record the height of you and your friends.

Name	Height (in inches or centimeters)

Construct a bar graph using the data above. Write a story about your graph.

2. Record the height and long-jump distances for you and your friends.

Name	Height (in inches)	Long Jump (in inches)

Construct a double bar graph using the data recorded above. Write a story about your graph.

APPENDIX 12

Raisin Experiment Activity Sheet

Name _____ Date _____

School _____ Class _____

Due: _____

1. Guess how many raisins are in your raisin box. *Guess:* _____

2. Record the guesses of all the students in your group in the table below.

3. What is the lowest guess? _____

4. What is the highest guess? _____

5. Which guess occurs most often? _____

6. Count the raisins in your box. How many are there? _____

7. Record the actual number of raisins in the boxes of all the students in your group. Use the table below.

8. What is the fewest number of raisins? _____

9. What is the greatest number of raisins? _____

10. How many raisins occurred most often? _____

Name of Student	Guess	Actual Count

11. Construct a double bar graph.

12. Answer the following:
    a. How many raisins did you have in your box? How close was it to your guess?
    b. Whose guess was the closest to the actual count?
    c. What brand of raisins did we use?
    d. What is another brand of raisins we could use? How do you think a box of the same size would compare?
    e. Think of a question that could be answered using your graph.

13. Write a story about the activity and your graph.

APPENDIX 13

Height over Time
Data Collection

Name _____ *Date* _____

School _____ *Class* _____

Due: _____

Keep a record of your height (or the height of a plant) for a period of sixteen weeks (or four weeks).

Date	Height (in cm)

Construct a line graph using the data above. Write a story about your graph.

APPENDIX 14

Temperature Data Collection Sheet

Name _____ Date _____

School _____ Class _____

Due: _____

Keep a record of the A.M. and P.M. temperature. Be sure to check the temperature at the same times each day during the next seven days.

Date	Temperature (in Fahrenheit)	
	A.M.	P.M.

1. Construct a line graph using either the A.M. or P.M. temperature.

2. Construct a multiple line graph using the data above. For each graph, write a short story.

APPENDIX 15

Sunrise/Sunset Data Collection Sheet

Name _____ Date _____

School _____ Class _____

Due: _____

Keep a record of the time of sunrise and sunset for the next fourteen days. You can obtain this information in the daily newspaper.

Day and Date	Time of Sunrise	Time of Sunset
Mean		
Median		

Construct a multiple line graph using the data above. Write a story about your graph.

APPENDIX 16

Picture Graph Activity Sheet

School _____   Name _____
Grade _____   Date _____

Graph 1. Picture Graph (1 symbol represents 1 student)

Graph Title: _____

Number of children in family

Legend:

Graph 2. Picture Graph (1 symbol represents 2 students)

Graph Title: _____

Number of children in family

Legend:

1. How many children in our class have 2 children in their families?

2. How many children are in the *most* number of families?

3. How many children are in the *fewest* number of families?

4. How are the two picture graphs the same?

5. How are the two picture graphs different?

85